The Neuroethics of Biomarkers

Oxford Series in Neuroscience, Law, and Philosophy

Series Editors
Lynn Nadel, Frederick Schauer, and Walter P. Sinnott-Armstrong

Conscious Will and Responsibility
Edited by Walter P. Sinnott-Armstrong and Lynn Nadel

Memory and Law
Edited by Lynn Nadel and Walter P. Sinnott-Armstrong

Neuroscience and Legal Responsibility
Edited by Nicole A. Vincent

Handbook on Psychopathy and Law
Edited by Kent A. Kiehl and Walter P. Sinnott-Armstrong

A Primer on Criminal Law and Neuroscience
Edited by Stephen J. Morse and Adina L. Roskies

Bioprediction, Biomarkers, and Bad Behavior
Edited by Ilina Singh, Walter P. Sinnott-Armstrong, and Julian Savulescu

Finding Consciousness: The Neuroscience, Ethics, and Law of Severe Brain Damage
Edited by Walter Sinnott-Armstrong

The Neuroethics of Biomarkers: What the Development of Bioprediction Means for Moral Responsibility, Justice, and the Nature of Mental Disorder
By Matthew L. Baum

The Neuroethics of Biomarkers

What the Development of Bioprediction Means for Moral Responsibility, Justice, and the Nature of Mental Disorder

Matthew L. Baum

OXFORD
UNIVERSITY PRESS

OXFORD
UNIVERSITY PRESS

Oxford University Press is a department of the University of Oxford. It furthers the University's objective of excellence in research, scholarship, and education by publishing worldwide.Oxford is a registered trade mark of Oxford University Press in the UK and certain other countries.

Published in the United States of America by Oxford University Press
198 Madison Avenue, New York, NY 10016, United States of America.

First Edition Published in 2016

Library of Congress Cataloging-in-Publication Data
Names: Baum, Matthew L., author.
Title: The neuroethics of biomarkers : what the development
of bioprediction means for moral responsibility, justice,
and the nature of mental disorder / by Matthew L. Baum.
Other titles: Series in neuroscience, law, and philosophy.
Description: Oxford ; New York : Oxford University Press, [2016] | Series:
Oxford series in neuroscience, law, and philosophy | Includes
bibliographical references and index.
Identifiers: LCCN 2015031170 | ISBN 9780190236267 (alk. paper)
Subjects: | MESH: Mental Disorders—diagnosis. | Bioethical Issues. |
Biological Markers. | Risk Assessment.
Classification: LCC RC455.4.B5 | NLM WM 141 | DDC 616.89/071—dc23
LC record available at http://lccn.loc.gov/2015031170

Foresight! Foresight!, which takes us ceaselessly beyond ourselves and often places us where we shall never arrive. This is the true source of all our miseries. What madness for a fleeting being like man always to look far into a future which comes so rarely and to neglect the present of which he is sure. It is a madness all the more destructive since it increases continuously with age; and old men, always distrustful, full of foresight, and miserly, prefer to deny themselves what is necessary today so as not to lack it a hundred years from now.

Rousseau (1979 [1762]), *Emile*, pp. 82–83

CONTENTS

LIST OF FIGURES AND TABLES

ACKNOWLEDGMENTS

I am indebted to a great number of individuals without whom this book would not have been possible. I am especially grateful to my Oxford DPhil advisors, Julian Savulescu and Mark Sheehan, as well as to Tony Hope, Neil Levy, Tom Douglas, Russell Perkins, Caroline Huang, Roger Crisp, Paul Troop, my two DPhil examiners, and the anonymous reviewers at Oxford University Press for providing invaluable feedback on the arguments in this book. Andrew McCall and Tad Baum generously commented on the book nearly in its entirety, which much improved this end product. Thanks to the District I Rhodes Scholarship Committee and the Rhodes Trust for taking a chance on a molecular neurobiology student who wished to pursue a degree in neuroethics, thus enabling this work. There are many who, through careful and kind administration, helped me navigate the formal hurdles necessary for the pursuit of the ideas described in this book: Christelle Kervella-Jones, Mary Eaton, Don Markwell, Piers Nye, Susan Barrington, Andrew Cooper, Glynis Price, Nailya Nikitenko, Miriam Wood, Deborah Sheehan, and of course Joan Bossert, Louis Giulino, Lynn Luecken and the production team at Oxford University Press. To the many others who have helped me along the way who are not by name acknowledged here—among them my colleagues in the Oxford Department of Public Health, the Ethox Center, the Uehiro Centre for Practical Ethics, Rhodes House, the Oxford Centre for Neuroethics, Balliol College, Harvard Medical School, and the Petrie Flom Centre at Harvard Law School—thank you, and I apologize for not acknowledging you individually. Finally, thank you to Emmilie Aveling for supporting my pursuit of this project despite its challenges.

Any errors, mistakes, weaknesses, and omissions that remain in this book after the help of this great number of people are entirely my own; for these errors, I apologize in advance.

The Neuroethics of Biomarkers

Introduction

1

The Biomedical Promise of Biomarkers

"See, it shakes a little," he said, holding out his hand. The man's fingers circled in unison with his index finger rubbing repetitively against the soft underside of his thumb; a classic "pill-rolling tremor," he went on to say. The man's tremor was the only visible symptom of his new diagnosis of Parkinson's disease. Parkinson's disease is thought to be caused by the deterioration of a certain population of neurons that produce the neurotransmitter dopamine, and are important for the control of voluntary movement. These neurons are progressively lost from a specific area of the brain called the *substantia nigra*, which is named after the characteristically dark pigment that the high concentration of dopamine gives these neurons. Interestingly, the paradigmatic symptoms of the disorder[1]—tremor, shaking arms, bobbing head, and slowness of voluntary movement—do not appear until the person has already lost *over 80 percent of these key neurons* (Nestler et al. 2009). Therefore, once a person's symptoms can be diagnosed as Parkinson's, this neuronal population is mostly dead, and treatment difficult.

Modern efforts in neuroscience are building an increasingly convincing case that many mental and neurological disorders, from this man's Parkinson's disease to mental retardation, schizophrenia, epilepsy, bipolar disorder, and Alzheimer's disease, *develop over time*. In what is sometimes referred to as a molecular cascade (Boenink 2009), these disorders are often preceded by increasingly aberrant molecular and circuit-level changes that develop over weeks, months, years, and even decades, *before* the appearance of the recognizable clinical symptoms of the disorders. As a student of molecular biology studying these disorders, I became increasingly fascinated by the possibility that many of the disorders as we currently diagnose them might actually be severe *end states* rather than newly onset dysfunctions. The idea is that with many disorders we are arriving very late, as if to the scene of a car accident; maybe things could be different if we could find the equivalent of sticky brakes, overinflated tires, and busted headlights that increase the likelihood of a crash.

I am not alone in that sentiment, as illustrated by the following excerpt from a paper authored by the former director of the National Institutes of Mental Health (NIMH), Tom Insel:

> Currently, mental disorders are diagnosed by symptoms that emerge at a late stage, presumably years after brain systems veer from more typical development. Diagnosing schizophrenia or bipolar disorder with the emergence of psychosis may be analogous to diagnosing coronary artery disease by myocardial infarction [heart attack]. One of the most hopeful approaches to reducing the morbidity and mortality of serious mental illness borrows a page from the cardiology playbook. By developing biomarkers [biological markers] for early diagnosis, we may be able to preempt many of the most disabling aspects of our most severe mental illnesses. (Insel 2009)

If we could identify this veering of brain systems earlier, would we have better success in treating the late stage? Could we intervene to delay or even prevent the onset of the disorder as we know it? Or could we at least prepare for the contingency, as we prepare for a possible hurricane?

The difficulty of adequately treating established disorders has prompted not only scientists, but also government, to establish as top priorities early detection and preventive medicine for brain-based disorders. In the United Kingdom (UK), the Foresight Report argued that preventive medicine in mental health is necessary to guard and maximize the nation's "mental capital" (Beddington et al. 2008). The UK Medical Research Council, in its 2010 review of mental health research, also called for greater emphasis on prevention in mental health (Sahakian et al. 2010).

For preventive medicine to be successful, however, we first need to advance methods of estimating who is at risk. In recognition of this necessity, a 2010 report from the Grand Challenges in Global Mental Health Initiative named the identification of biomarkers of increased risk of mental disorder as "Goal A" (Collins et al. 2011).

Clearly, many big players in the public sector are convinced that the development of predictive biomarkers holds vast potential for changing the health of individuals and populations, but what else might the power of bioprediction change? The following fictional, but reality-based, cases illustrate four potential domains of change.

First, the development of predictive biomarkers is poised to unsettle how we relate to the definitional categories for which the biomarkers are developed. Consider "Risky Definitions."

RISKY DEFINITIONS

> *Rick has been bothered by the feeling that people on the city bus he rides to school are talking about him behind his back; he knows they probably are not but he can't shake that uncomfortable feeling and he thinks this is starting to interfere with his performance in school and sports. He seeks a psychiatric consultation and after a battery of testing is told that he has a condition sometimes called "psychosis risk syndrome": in experimental settings 20–40 percent of young people who scored as he did on these tests go on to have a psychotic episode in the next two years. When Rick's primary care physician (PCP) learns of this consultation, he is displeased, explaining that since the majority of people will not go on to develop psychosis, the category needlessly causes fear and stigma.*

Inclusion in the fifth edition of the *Diagnostic and Statistical Manual of Mental Disorders* (*DSM-5*) of such a category based on risk of psychosis was hotly contested and eventually was omitted amidst concerns like those voiced by Rick's PCP: that we should not have disorders *based on risk* because many of those individuals in the category will not go on to develop the thing they are at risk for (as discussed in Chapter 5; also see Singh and Sinnott-Armstrong 2014). With the development of biomarkers, such contentious debates on risk categories are likely to multiply. What criteria can we look to in evaluating these risky disorders?

Second, the development of predictive biomarkers may change the extent of the obligations we owe to others; consider "Risky Driving."

RISKY DRIVING

> *Sally, a 33-year-old female, was brought into the emergency room after being hit by a bicycle as she crossed the street. Sally had a few scrapes and bruises, but was otherwise unharmed. Having had an uncle pass away unexpectedly from a heart attack, and feeling pain in her chest, Sally was quite distressed about her heart. An echocardiogram (an ultrasound on the heart) was performed. The echo-technician was surprised to find that the walls of Sally's heart were profoundly thickened. When Sally learned that the presence of this biomarker, increased heart wall thickness, corresponds to an increased risk of sudden cardiac death, the first thing she did was to make arrangements to stop driving the van for her children's playgroup. When asked about that decision, Sally explained that she felt "it would have been irresponsible to do otherwise."*

Sally felt that the knowledge of her biological risk required her to take special precautions to minimize the likelihood that the risk would translate to harming those she cared for. Under what conditions should knowledge of biological risks, as estimated by biomarkers, enhance moral responsibility in this way? Might there be some cases in which one could be blameworthy for continuing to drive in ignorance of the risks one might pose to others?

Third, the development of predictive biomarkers may change those whom we punish, and how much we punish them; consider "Risky Punishment."

RISKY PUNISHMENT

In Brad's murder trial, his lawyer argued for a reduction in sentence on account of Brad's genetic, environmental, behavioral, and neuroimaging markers, which research has shown are enriched in groups of individuals that, on the whole, are more likely to commit reactive violence. The opposing legal team objects on the grounds that data gathered on groups do not validly apply to Brad as an individual, that these biomarkers are problematically probabilistic, and that a proneness to violence is no grounds for a reduction in legal responsibility.

As biomarkers risk flooding the courts, there are split opinions about whether and under what conditions biomarkers should rightly reduce legal responsibility. While the courts have structures for determining the relevance of a given mental disorder, it is unclear where to fit biological risks.

Fourth, biomarkers may change the obligations of society to individuals in terms of the way it chooses to distribute resources or opportunities; consider "Risky Allocation."

RISKY ALLOCATION

The social service of Hynkley is reviewing the way it currently allocates its resources for at-risk children. Among the proposals being considered is one that controversially suggests that resource allocation should be prioritized through violence risk estimation, incorporating biomarkers. It is argued that children at higher risk of future violence are more likely to have life opportunities curtailed by the justice system, and that the social service should prioritize these worst-off children in the distribution of the service's limited resources.

Are those with higher risk of something bad in the future actually worse off? After all, most of the children in the hypothetical village of Hynkley will not end up in trouble with the justice system, just like most of the young people with "psychosis risk syndrome" will not develop psychosis. Are such risky assessments plausible grounds to change how we should distribute resources? Just as public health is expanding to consider social influences on health, should institutions of distributive justice also expand to consider these biological influences on social outcomes?

Though this science is young, the rapidly expanding development of biomarkers in neuroscience will increasingly enable the estimation of the likelihood (bioprediction) of future adverse psychological events, from the emergence of full psychotic episodes and the onset of dementia to impulsive violent reaction.[2] The medical potential of biomarker discovery is immense and worthy of the excitement that surrounds it, but the discussion needs to extend to the ethical implications of bioprediction (the use of those biomarkers), some of which are illustrated by "Risky Definitions," "Risky Driving," "Risky Punishment," and "Risky Allocation" and will be discussed as this monograph unfolds. As shown in Figure 1.1, discussions on what we *ought to do* with the biomarker explosion lag far behind discussions of what we *can or might do*. Though discussion of the neuroethics of biomarkers is just beginning, it is not contained solely within the vials of the medical clinic, but spills into courts and statehouses, drawing the interest of a uniquely interdisciplinary group of scientists, lawyers, doctors, and philosophers.

It is the preliminary examination of several ethical issues of bioprediction, and the construction of a philosophical groundwork with which to assess others, that are the goals of this book. What makes bioprediction interesting is that it not only raises moral challenges, but also challenges our moral frameworks. I argue that much of the current ethical controversy about biomarkers stems from disagreements about how (or whether) to integrate bioprediction's explicitly probabilistic predictive information into existing medical, legal, and political structures.

Much existing debate centers on a perceived categorical division between disorders and biological risks. The difficulty is in part that our institutions have gotten on well *enough* by cleaving the world into the healthy and the sick. Consequently, medicine, law, and society have special protocols for disorders, but not *risk of disorders*. A disorder changes how we allocate resources, like health care or social services. A disorder changes the moral attitudes we hold, like whether we excuse or blame—and how much. Or a disorder creates new responsibilities for the person who has it—for example, driving restrictions if one has epilepsy or visual impairment.

But what do we do if someone does not have a named disorder, but has a biological risk?

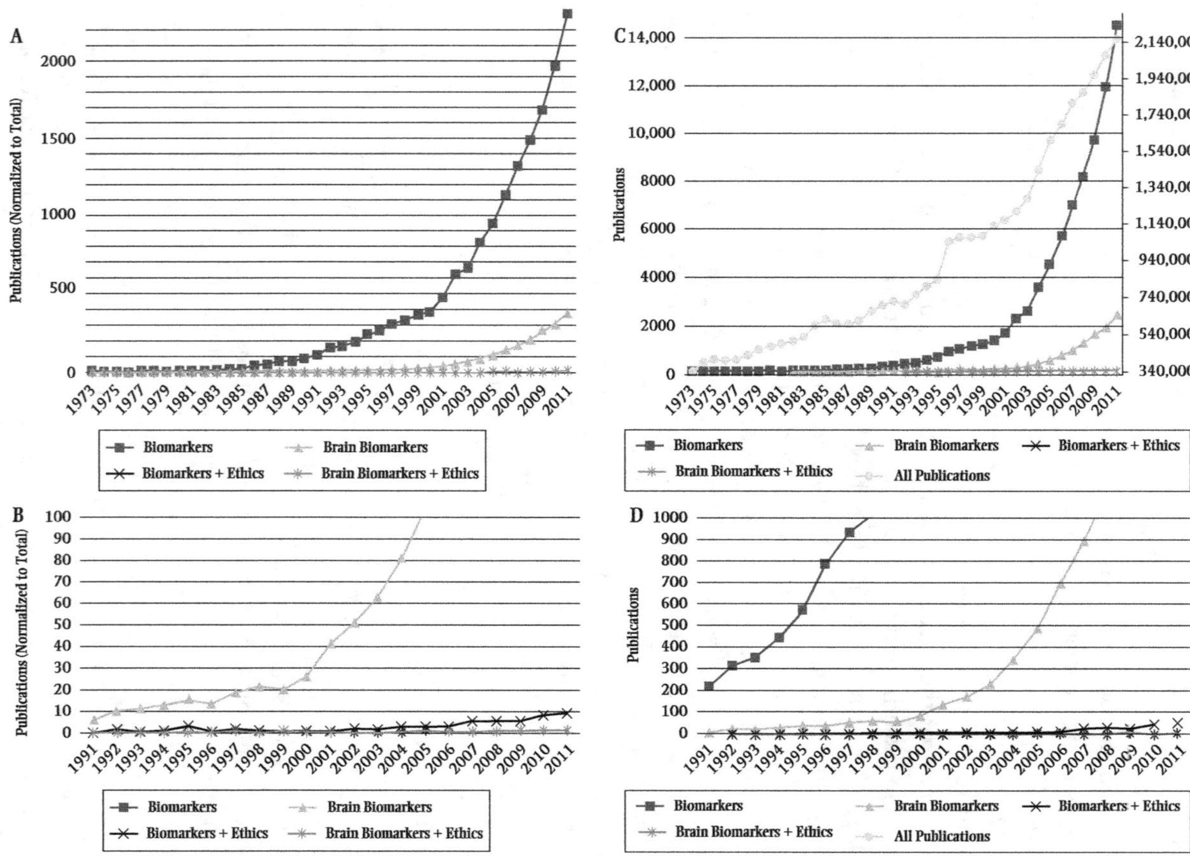

Figure 1.1 Continued

Figure 1.1 *Discussion of the ethics of biomarkers lags far behind the science.* A systematic search in SCOPUS suggests that publications per year on biomarkers have exhibited exponential growth in the past two decades. Publications on biomarkers related to brain/behavior show a similar, but more recent, exponential increase. The biomarkers papers that mention ethics or morality, in contrast, have appeared only in the last few years and show a linear increase. When one looks at papers on biomarkers related to brain/behavior that also mention ethics or morality, one sees that there has been only a handful of publications per year, and little growth. When one considers that these measures count all biomarkers publications that even mention ethics or morality in the abstract, it is clear that an extended ethical discussion is both timely and needed. A and B show publications per year normalized to total publications (containing the word "the" in the abstract), with B showing a rescaled axis to better visualize ethics papers. C and D are absolute numbers of publications per year, with D rescaled. The right-hand Y-axis in C is all publications.

A discussion of the neuroethics of biomarkers is likely to be plagued by this tension until it is recognized that the problem lies not with the imperfect probabilistic quality of bioprediction *per se* but with the flawed assumption, often made, that disorder is categorical (sick vs. well). This binary conception is especially strong surrounding disorders in psychiatry (and present, though less strongly, in neurology) and so my argument will address brain-based disorders at length.

Through the extended dialogue of this monograph I aim to show how bioprediction (1) exposes this flaw in one of the most central tenets of medicine upon which our moral frameworks depend (the nature of disorder), (2) enables the recognition that risk (and magnitude of risk) is the most integral part of the concept of disorder, and (3) points the way toward a reorientation of the concept of disorder through which predictive biomarkers can act synergistically, rather than antagonistically, with our existing moral structures. By arguing for the abandonment of this no longer defensible binary in favor of the probability-based concept of disorder that I develop, I hope that this work can add sufficient conceptual clarity to meaningfully advance debates concerning the most controversial ethical implications of predictive biomarkers.

The potential practical benefits of applying biomarkers to medicine, public health, and other social institutions are driving the development of biomarkers; few have considered, however, that the development of biomarkers may also shift the philosophical assumptions at the foundation of these institutions.

After developing and defending my risk-reoriented concept of disorder, which I call the Probability Dysfunction, I demonstrate its usefulness by applying it to many of the ethical issues raised by "Risky Definitions," "Risky Driving," "Risky Punishment," and "Risky Allocation": What are the implications of biomarkers for the definition of mental disorders in the *DSM-5*? For "risk" disorders of psychosis or dementia? Might biomarkers not only narrow but—counterintuitively—drastically extend the range of acts and omissions for which we might be held responsible? When does a biopredicted risk become morally significant? How do biomarkers change our obligations of distributive justice?

There are several other distinguishing features of this monograph. While existing literature trends toward a reductionist approach to bioprediction (i.e., considering the implications of single biomarkers in isolation), I consider the strongest ethical challenges to come from combinations of many biomarkers (genetic and non-genetic) into bio-actuarial tools similar to augmented versions of a FRAX fracture risk score or Reynold Risk Score for cardiovascular event. With the advent and increasing power of high-throughput "omics" technologies (outlined in Sung et al. [2012]), and the coevolution of statistical and machine-learning techniques like principle component analysis and support vector machines that are able to find patterns in many small variations across large data sets, biomarker discovery and assembly into bio-actuarial tools are poised to proceed at an unprecedented pace. By emphasizing bioprediction's usefulness for enabling moral agency and distributive justice, moreover, I challenge a commonly asserted notion that the acceptability of bioprediction turns primarily on whether it enables better medical treatment.

THE ORGANIZATION OF THIS BOOK

This book is organized in an introduction and three subsequent parts. In the remainder of the introduction, I set necessary definitional groundwork in the form of a taxonomy of biomarkers and discussion of some of the many ways in which biomarkers might be used; for example, the same biomarkers might meaningfully be used for both bioprediction and retrodiction, depending on one's frame of interest. In Part I, I will (a) highlight challenges that the development of predictive biomarkers raises for brain disorder, (b) propose to address these challenges with a novel bio-actuarial model of disorder based on risk of harm, and (c) demonstrate the immediate benefits of this new model by using it to clarify and make practical recommendations for the nosological debates surrounding "prodromal psychosis" and "prodromal Alzheimer's disease." In Parts II and III, I will show how this reorientation, first, illuminates implications of predictive biomarkers of behavior for individual moral responsibility, and, second, together with prioritarian and egalitarian concerns for the worst off, demonstrates an obligation to further develop and provide access to biopredictive technologies. Part II focuses on the implications for moral responsibility of two specific applications of bioprediction: the prediction of epileptic seizures in traffic safety, and the MAOA gene x environment predisposition for impulsive aggression. These examples demonstrate the potential for bioprediction to both decrease and enhance the moral responsibility of individual agents, and also illustrate the conceptual advantages of my new definition of disorder in these discussions. Part III outlines how select social services might alter their resource allocation in light of this new technology, and through this discussion illuminates and defends the most counterintuitive results of my new definition: that prioritarian or egalitarian concerns for the worse off create a moral obligation to invest resources in the further development of biopredictive technologies, and to facilitate access to these technologies.

This book is intended for an interdisciplinary readership of physicians, scientists, law professionals, and philosophers. Because this broad group will have an equally broad distribution of preferences (or tolerances) for theoretical versus practical issues, I should say a few words about how to read this book. The Introduction (Chapters 1 and 2) is recommended for all readers, as it provides background on biomarkers necessary in understanding both the theoretical chapters (3 and 4) and the chapters that deal directly with practical issues

(5–8). Outliers toward the practical side of the preference distribution may be most interested in the discussion in Chapters 5–8. Chapters 3 and (especially) 4 weigh heaviest with theory, as it is here where I construct and defend my theoretical reconceptualization of the concept of disorder around risk. While I hope that most readers will be open to this deep theoretical engagement—I firmly think that it provides a useful high-level framework for approaching the applied issues in the neuroethics of biomarkers, both the ones in this book and generally—I have tried to structure Chapters 5–8 such that those practical discussions will still be approachable and meaningful if one reads only the thin version of the risk reorientation of disorder that can be found in the introduction to Part I, which precedes Chapter 3. In the instantiation of such a structure, I sometimes use slightly different tactics to arrive in Chapters 5–8 at some of the same conclusions from the theory of Chapters 3 and 4, but I am optimistic that those who do engage directly with the theory will find these complementary approaches helpful rather than redundant.

Let us now begin to clearly define the terminology of bioprediction, and the scope of the argument; this is the goal of Chapter 2.

NOTES

1. Throughout this book, I will use "disorder" as a broad term to encompass what some alternatively call a disease, illness, injury, or malady, except when using the proper name of a specific disorder (e.g., Parkinson's disease). This use matches Wakefield (2007) and is discussed in Part I.
2. Part of the ethical complexity surrounding bioprediction is that many predictive biomarkers are emerging indirectly—not from science geared specifically at prediction, but rather from more basic science; thus the standards and goals of those developing biomarkers risks diverging from those (e.g., the Department of Motor Vehicles) who may wish to use them for bioprediction. Much of this indirect discovery arises because biomarkers act as essential footholds into investigating the early causes of disorder. If there is a disorder about which little is known, a first step is to look for a gene variation, a pathway disruption, or another biological variation with which the disorder of interest is associated; that biological variation can then be modeled (recreated experimentally) and further investigated in animals. Within the field of brain disorders, where the behavioral and emotional phenotypes are often difficult to capture precisely in animals, this strategy of modeling biomarkers of a disorder is especially important (Sarnyai et al. 2011; Malkesman et al. 2009). Results from animal models can both feed-back to guide better biomarker discovery and feed-forward to aid development of potential treatments or prevention. The discovery and improvement of biomarkers are integral to the promise of molecular medicine, the rational design of treatments based on the molecular dysfunction undergirding the disorder. The goal is to take the biological insight gained at the laboratory bench back to the patient's bedside.

2

Bioprediction of Brain Disorder

Definitions and Scope

I have already used the intimately related terms "bioprediction" and "biomarkers," but a formal statement of their relationship is necessary.

> *Bioprediction*: the use of a biomarker to predict the likelihood of something occurring in the future.

The literature often under-describes the definitions of biomarkers and bioprediction, and as such it is not uncommon to encounter very educated professionals who hold very different assumptions about what, in fact, a biomarker is. In this chapter I will provide sufficient background on the types and properties of biomarkers to avoid these common confusions, which will enable the clear articulation of my reoriented model of disorder in Part I. (See Table 2.2 at the end of this chapter for summary of these definitions other useful terms.)

WHAT IS A BIOMARKER?

In 2001, the National Institutes of Health conducted a working group to standardize the definition of biomarker, and I will use this standard definition:

> *Biological marker (biomarker)*: A characteristic that is objectively measured and evaluated as an indicator of normal biological processes, pathogenic processes, or pharmacologic responses to a therapeutic intervention. (Atkinson et al. 2001)

One should observe that this definition is very broad. Three especially famous examples of biomarkers are blood pressure, cholesterol levels, and blood glucose levels. Two other biomarkers in popular awareness are HIV viral loads (number of viruses) and white blood cell counts as measures of diagnosis, prognosis, and treatment response in HIV infection.[1] A key difference between biomarkers and classical signs and symptoms (like the pill-rolling tremor) is that they are objectively measured.

Biomarkers, however, can come in many shapes and sizes. They can be simple genetic variants, gene x environment interactions, altered mRNA expression, protein expression and modification, lipid expression, metabolite levels, cellular function, brain structure, brain function, and electrical signal patterns. This is not an exhaustive list, and indeed all of the "-omics" technologies described in Table 2.1 can discover biomarkers. If traditional cellular pathology (changes in the appearance of cells or tissues) can be observed (with the help of a stain) and objectively measured, for example by quantitative image analysis, this too might be conceived through the paradigm of biomarkers.[2]

Biomarkers are sometimes further divided into "wet" and "dry" (Filiou & Turck 2011). "Wet" biomarkers involve a bodily fluid such as blood, urine, saliva, or cerebrospinal fluid surrounding the brain and spine. "Dry" biomarkers include imaging techniques such as positron emission tomography (PET), single positron emission computed tomography (SPECT), electroencephalogram (EEG), magetoencephalogram (MEG), magnetic resonance imaging (MRI), and functional MRI (fMRI).

A biomarker is often (but not always) a *quantitative* measure. This, too, differentiates biomarkers from classical signs and symptoms. How high is the blood pressure? How much epileptiform activity? How many viruses? To what

Table 2.1 TYPES OF BIOMARKER 'OMICS

Type of -omic	**Brief Description**
Genomics	The primary sequence of DNA (Bloss et al. 2011)
Exomics	The primary sequence of DNA that is thought to code for proteins (Klassen et al. 2011)
Transcriptomics	The primary sequence of mRNA transcribed from genes (Karsten et al. 2011)
Proteomics	One's proteins (Martins-de-Souza, Guest, et al. 2011) and post-translational modifications of proteins (such as phosphorylation) (Jaros et al. 2012)
Connectomics	The connections between neurons in the brain (Alivisatos et al. 2012; Reid 2012)
Lipidomics	The profile of lipids in a cell population (Quehenberger & Dennis 2011; Wood 2012)
Metabolomics	The profile of molecular metabolites (Hyötyläinen 2012)
Epigenomics	Modifications of DNA such as methylation states (Zhou et al. 2011)

degree has the brain area shrunk? A biomarker, however, can also be an all-or-nothing indicator, such as the presence or absence of a risk-associated genetic variant. Even these all-or-nothing indicators will often have a quantitative aspect, as is the case where one copy of the APOE ε4 gene variant leads to a quantitative percentage increase in risk of Alzheimer's disease, and possession of two copies further increases that risk (Corder et al. 1993).

As we shall see, it is this shift toward quantification, and the resultant spectrum of human variation to which biomarkers draw attention, that pose the greatest challenge to existing binary definitions of disorder and institutions reliant on that definition.

A few functional categories of biomarkers are needed to help parse the philosophically relevant aspects of their use: whether they are predictive, real-time, or retrospective.

> *Predictive biomarker*: a biomarker used to estimate the likelihood of something happening in the future.
>
> *Real-time biomarker*: a biomarker used to estimate the likelihood that something is currently happening.
>
> *Retrospective biomarker*: a biomarker used to estimate the likelihood that something happened in the past.

A single biomarker might be predictive, real-time, retrospective, or all three at once, such as impeded contrast-agent flow on an angiogram of a blood vessel in the motor cortex of a person's brain. It is a predictive biomarker of the cognitive, emotional, or behavioral symptoms that person is likely to experience in the future, a real-time biomarker of whether the surrounding brain tissue is experiencing a toxic lack of oxygen, and a retrospective biomarker of whether a blood clot or atherosclerotic plaque formed elsewhere in the body, traveled to the current site, and lodged in the blood vessel. Likewise, very low blood glucose level might be a predictive biomarker of an adverse event (the symptoms of a hypoglycemic state), a real-time biomarker of a patient's metabolic processes, and a retrospective biomarker that the person received too much insulin.

By using this definition of a predictive biomarker, we can simplify the definition of bioprediction as follows:

> *Bioprediction*: the use of a predictive biomarker.

Much of the shift toward bioprediction comes when biological markers that have previously been measurable only after an event are now detectable, due to a technological advance, before that event. For example, pathologists have long examined the appearance of organs and the microscopic structure of tissues taken from dead patients to develop hypotheses explaining what led to the patient's symptoms. This construction of a hypothesis based on the appearance of tissue stained or treated in particular ways is often called a clinical-pathological correlation. The advent of technology that enables the pre-mortem imaging of the pathologist's clues can bring the clinical-pathological correlation forward in time, a phenomenon which reveals the intimate relationship between bioprediction and the field of pathology. This is currently happening for Alzheimer's disease.

Pathologists hypothesized a particular degenerative process that may be involved in the cognitive dysfunction and deaths in what is currently called Alzheimer's disease after noticing that a special silver stain of dead brain tissue enabled the visualization of two distinct tissue-staining patterns (biomarkers) that were named plaques and tangles. With the recent development of several small molecules that, when injected into the blood, flow across into the brain, transiently bind to the proteins that clump together to form plaques and tangles, and then emit a small amount of radiation in the form of positrons that can be detected by noninvasive brain imaging (PET), the burden of these biomarkers can now be estimated in living people, many of whom have little or no cognitive impairment (Ewers et al. 2011; Jack & Holtzman 2013). Thus, the presence of plaques in the dead patient's brain is a retrospective biomarker that, now detectable in a live patient, has become a predictive biomarker. I will return to the bioprediction of Alzheimer's disease in Part I, Chapter 5.

As this book is concerned with bioprediction, discussions will primarily deal with predictive biomarkers, but the potential for predictive biomarkers to function also in real-time or retroactively will be relevant to an ethical and philosophical discussion of bioprediction. In Part II's discussion of biomarkers in a legal context, it is relevant that a biomarker could have performed a predictive function in the past, even though it functions retrospectively in court. For example, an ethical analysis of a trial for violent crime must take into account that a biomarker could have predicted that the defendant had an increased likelihood of violent response in a certain context.

In science, and in some realms of philosophy, a great deal of effort is expended to differentiate between causes and correlations of a given phenomenon. In the medical sciences, this effort is usually justified by the higher aim of finding ways to interfere in that causal chain (treatment); in philosophy, as I shall show in Part I, Chapter 4, causation can be relevant to the conceptual analysis of what defines a disorder. The term "biomarker"—even "predictive biomarker"—does not differentiate between causation and correlation. Only in specific cases does one need to differentiate between causation and correlation, however, and even then, usually with broad strokes.

Predictive biomarkers can further be subdivided, therefore, into causative or correlative, keeping in mind that in complex processes, there are often many partial causes. We might gather data that an increase in a biomarker level (event B) is often followed by a certain phenomenon (event C). Though B in this case is a predictive biomarker of C, we might be unsure of whether B caused C or whether some third unknown process, A, caused both B and C. For example, consider the following quotation from a 2012 paper in *Nature* (Straussman et al. 2012) investigating resistance of cancer cells to a specific chemotherapy (the specific acronyms are unimportant):

> To establish HGF [B] as the cause of drug resistance [C] and not simply a biomarker for resistance [i.e., to establish B → C rather than A → B and A → C] we tested the ability of recombinant HGF to induce resistance, as well as the ability of HGF-neutralizing antibodies or the MET-inhibitory small-molecule crizotinib to block fibroblast induced PLX4720 resistance.[3] (Straussman et al. 2012)

In this study, the researchers recreated the altered HGF level [B] to see if that alone could cause the resistance [C] (whether B was sufficient for C). Then, they blocked the HGF-increase to see whether B was necessary for C. Just like all squares are rectangles, but not all rectangles are squares, all measurable biological causes are predictive biomarkers if measured, but not all predictive biomarkers are causes.

If we seek to interfere in the causal chain, we must further recognize that some causative biomarkers, if modified, reduce the risk of the future event, but others, even if modified, may not reduce the risk of the future event (i.e., if the causative process is not reversible). This latter type of result is particularly common when a condition *develops over time.*

Whether the biomarker is a cause or a correlate is often irrelevant to its usefulness. Of course, this statement sounds blasphemous considering the fact that the importance of differentiating between causation and correlation is drilled *ad nauseum* into the minds of undergraduates. If all that we want to do is to *predict the event*, C, however, then we should not care if the biomarker is causative or not. Often, we have *some other way* of interfering in the causal chain that is *independent of the biomarker.* If my bones ache before a heavy rain (and I want to predict rain), I could not care less that my aching bones are not causing it to rain. All I care about is that the biomarker is sensitive enough (when my bones ache, it is more likely that it will rain) and that the biomarker is specific enough (my bones ache less often when it is not going to rain than when it is going to rain) for my desired purpose. This type of biomarker is good enough if all I want is to plan something around that information (should I bring an umbrella?). This property of predictive biomarkers is important when we consider seizure prediction and structurally similar cases (Part II, Chapter 6), where the availability of bioprediction may give reason to take special precautions to minimize harm to others (in this case, by modifying driving behavior), even if bioprediction does not improve medical treatment.

From a practical perspective, only if we desire to manipulate the biomarker directly to avoid the event, C, do we need to make the distinction between causative and non-causative biomarkers. If I want to *stop it from raining*, I want to know if taking some ibuprofen to stop my bones from aching will do the trick (or if I preemptively take ibuprofen even when my bones don't ache). This division in goals between those who merely want to predict something and those who want to *know how to intervene* is one point of tension when it comes to the neuroethics of biomarkers; scientists and clinicians, who often develop biomarkers for therapeutic purposes (i.e., intervention), may have very different biases as to when these predictors are "ready for prime time" than lawyers or laypersons who may be interested in improving predictions alone.

A major force in the discovery of biomarkers is the search for a subdivision of causal biomarkers called "surrogate endpoints." If the manipulation of a biomarker changes the likelihood of the event it predicts in a reliable way under defined conditions, then it may qualify as a "surrogate endpoint," a concept that has largely been developed in the context of pharmaceutical trials. In pharmaceutical trials, a surrogate endpoint is a subset of predictive biomarker that is causally linked to a "clinical endpoint" in such a way that a decrease in the rate of the biomarker indicates a decrease in the rate of the clinical endpoint (and an increase of biomarker indicates an increase in likelihood of the clinical endpoint). High blood pressure is an example of a surrogate endpoint for heart attack. The US Food and Drug Administration (FDA) can grant approval to pharmaceuticals based on clinical trials using surrogate endpoints under the following conditions:

> [. . .] the drug product has an effect on a *surrogate endpoint that is reasonably likely, based on epidemiologic, therapeutic, pathophysiologic, or other evidence, to predict clinical benefit or on the basis of an effect on a clinical* endpoint other than survival or irreversible morbidity. (my emphasis, quoted in Atkinson et al. [2001])

This FDA policy is evidence of how modern medical practice nests surrogate endpoints within each other like a set of Russian nesting dolls. The FDA considers all non-"survival or irreversible morbidity" endpoints, for example, as surrogate endpoints. By this reasoning, stroke occurrence is a surrogate endpoint that probabilistically predicts survival or irreversible morbidity sometime in the future (the clinical endpoints). Blood pressure, in turn, can act as a surrogate endpoint for stroke, in that an intervention that increases or lowers blood pressure will increase or lower risk of stroke (as well as "congestive heart failure, and subsets of cardiovascular death") (Atkinson et al. 2001). Once we recognize the probabilistic relationships between the sequential nesting dolls, we have a common statistical currency to move from one level to another; this will be central in my re-conceptualization of disorder as a series of risk bands of a kind of harm (the non-surrogate endpoints).

There is a dis-analogy in the Russian nesting dolls, however, as the relationship between predictive biomarkers and the more advanced stages they predict, rather than being one-to-one, is often multi-forked. High blood pressure is not only a predictive biomarker of stroke, for example, but also of heart attack and of a range of other outcomes. Biomarkers often have significant *pleiotropy*, which means that they contain probabilistic information on a *range of possible future events or conditions* rather than just one event. This pleiotropy, as we shall see, is a major contributor to the need to move away from a binary concept of disorder into one that is probability-based.

When it comes to prediction of a specific future condition, most single biomarkers identify small amounts of increased risk. That is, the predictive value of most single biomarkers is very small, particularly for complex disorders, including most mental and neurological disorders. As a result, most fall into the category of "imperfect," rather than "perfect," biopredictor. A perfect biopredictor would, with 100 percent certainty, indicate the future onset of a particular condition, whereas an imperfect predictor shows that the risk is heightened with respect to the population average. Sometimes, in debates involving the intersection of biomarkers and ethics, people claim that the fact that a certain biopredictor is not a "perfect" biopredictor means that it is of no value to the ethical or legal debate. We will encounter this argument in the discussion of how predictive biomarkers may, in certain cases, lead to a reduction in moral responsibility in Part II, Chapter 7. Once we relax our desire to think in binary and begin to see the world probabilistically, it will be clear that this type of certainty cannot be a requirement for relevance in these debates.

MULTIPLEXING: COMBINING MULTIPLE SMALL CHANGES ACROSS A DATA SET TO INCREASE PREDICTIVE POWER

While 100 percent certainty in prediction cannot be a requirement for relevance to ethical debates, the magnitude of the predictive probability may indeed matter. It is not uncommon to encounter skepticism about whether biomarkers will have anything but the weakest predictive value, and even more skepticism about whether predictive values will reach levels at which they can be practically useful, especially when it comes to guiding policy or law (see Wolpe 2014). It is important, therefore, to discuss several strategies that have been employed to increase the predictive power of biomarkers.

One strategy is illustrated within human genetics: some argue that while looking for common factors of a disorder will yield markers of small effect size, looking for very rare genetic variants (many of which are *de novo*, or "personal") may yield markers of large effect size. This strategy has been moderately successful in research on mental impairment and epilepsy, and some scientists posit that rare variants of larger effect might also be players in disorders like schizophrenia (Mitchell 2011; Mitchell et al. 2011; Sebat et al. 2009). However, even the rare genetic copy number variants (variations in how many copies of gene regions one possesses) that have been found confer odds ratios of 1.5–10 (i.e., that the odds of possessing the rare variant and developing schizophrenia are 1.5 to 10 times the odds of not having the variant and developing schizophrenia, which roughly would be moving from a general incidence of 1 percent to 1.5 percent or 10 percent) (Sebat et al. 2009). This is a big change, but still low in terms of absolute predictive value.

A second strategy is to multiplex biomarkers, that is, to collect a whole cadre of biomarkers simultaneously, and then to create algorithms that cobble together subpopulations of biomarkers to capture the highest amount of risk information; this strategy has grown increasingly popular in light of the growing consensus that most common disease categories are heterogeneous or otherwise biologically complex (Bahn et al. 2011). For example, rather than looking for changes in size or activity of an individual brain area on a brain scan, a team might use multivariate analysis to look for a pattern of changes across the entire brain. Or, instead of assaying the predictive power of the level of a single fluid protein, a group might measure a large and varied collection of metabolites and proteins and use machine learning or "greedy" algorithms that sort through patterns of variation *across all of the markers collected* to end up with a group of seemingly unrelated markers, variation in which explains a larger difference between control and disorder than any marker alone.

A research team recently applied this latter strategy, for example, to move towards better estimates of risk of psychosis (Perkins et al. 2015). They used a multiplexed immunoassay to measure 151 biomarkers related to inflammation, cellular growth, oxidative stress, cellular metabolism, and hormone signaling in blood samples from help-seeking young people identified behaviorally to be at high risk of psychosis (~30 percent risk of a psychotic episode within 2 years; see Chapter 5 for discussion of the category) and who had enrolled in the North American Prodrome Longitudinal Study (NAPLS 2). Using a "greedy" algorithm, the team was able to identify 15 biomarkers that could be applied as an index to classify a subgroup of help-seeking young people at even higher risk of psychosis; 72 percent in this subgroup went on to have a psychotic episode compared to ~30 percent in the risk group identified with clinical evaluation alone. Though this result needs to be validated in an independent "test" cohort, it forecasts the potential multiplexing has to drastically increase the predictive power of biomarkers.

Much existing work has focused on multiplexing biomarkers within a single modality (e.g., variation in genes, blood analytes, or aspects of brain images), and there may be further potential in combining sets of both wet and dry biomarkers. A second group within the NAPLS 2 study, for example, has shown that the rate of thinning of certain cortical areas is increased in those who go on to develop psychosis compared to those at risk who do not, or to control subjects, and that this thinning could be predicted by inflammatory cytokine levels at baseline (Cannon et al. 2015; Chung & Cannon 2015). Though this research is still young, we may in the not-so-distant future develop a combination of wet and dry biomarkers that can reproducibly identify groups of individuals with very high probabilities of having a first psychotic episode—before it occurs.

Combining multiple modalities of biomarkers is already the reality within dementia risk estimation. Several clinical trials are at present measuring a combination of multiple fluid biomarkers and imaging biomarkers, placing dementia risk-estimation far in advance of psychosis risk-estimation (Jack & Holtzman 2013). We will return to a discussion of high-risk categories of psychosis and cognitive impairment in the reorientation of the concept of disorder (Part I, Chapter 3).

The take-home point about multiplexing, for the purposes of this book, is that by analyzing *sets of biomarkers* rather than individual biomarkers, it may soon become possible to reach levels of predictive value previously thought to be unachievable. This will become increasingly true as these multiplexed assays combine with other "Big Data" tools that incorporate not only biomarkers but traditional signs, symptoms, and behavioral data into bio-actuarial algorithms. Discussions of the ethical and philosophical implications of bioprediction, therefore, should consider that the small predictive values that have been the rule in the genetics of complex disorder may not hold true for biomarkers when combined into multiplexed assays and Big Data approaches; rather, the probabilities of certain events may approach and *even exceed the predictive value of traditional diagnoses*. If it is the magnitude of probabilities that matters to the ethical debate, as I shall argue is often the case, then the development of multiplexing biomarkers represents a significant reorganization of the bioethical landscape.

Biomarkers and Other Sources of Risk Information

As I alluded in the preceding paragraphs, I think it likely that biomarkers will be combined with other kinds of information from which risk estimates might be gained, for example traditional signs or symptoms, behavioral markers, epidemiology, or markers that can only be subjectively measured. By restricting their biomarker study to the adolescents identified to be at high risk of psychosis through a set of behaviors and clinical measures, the NAPLS team is essentially trying find a combination of behavioral/clinical markers and biomarkers with improved predictive power. In Chapter 4, I highlight that a major function of grouping people into traditional diagnostic categories using familiar signs and symptoms is to estimate the risk of future harm to individuals within that group; thus diagnostic categories with criteria composed of both objectively measured biological markers and more subjective markers such as symptoms are examples of this type of combined prediction. One objective of this book, therefore, is to prompt an integrative approach to risk information, of which biomarkers are an increasingly abundant source. Where there are radical departures between what our moral and legal structures require of us in relation to different categories of risk (e.g., traditional diagnostic categories *vs.* risk groups defined by biomarkers alone), we should be critical about the basis upon which those categories are claimed to be meaningfully different.

Sometimes, however, we may think it justified to approach different sources of risk information differently. In Chapter 4, for example, I argue that one such distinction may be when a marker is associated with increased risk of harm *primarily because of an injustice in how a society is structured*; if a biological variation is associated with risk of harm primarily because a community is violent and prejudiced, this might rightly change the kind of obligations we have toward that individual.

We may also find that there are quantitative differences between biomarkers and other sources of risk information. For example, biomarkers may be more likely to be combined successfully into large computational models with high predictive power, precisely because they can be more objectively measured and standardized across multiple sites and large cohorts of people. Biomarkers might be more likely to be assayable well in advance of the harmful biological event (e.g., psychotic episode or seizure). Finally, biomarkers may be more likely to be nakedly probabilistic. And because this nakedness is less easily clothed in the binary thinking that obscures the probabilistic risk information within diagnoses and some subjective measures, biomarkers can brilliantly illuminate the dark reaches of our moral and legal structures that should be revised to more systematically—and fairly—handle probabilistic states.

A Note on Genetic Versus Non-Genetic Predictive Biomarkers

My focus in this book is on *non-genetic* biomarkers. I adopt this focus not because I think that genetic biomarkers are unimportant or because I think they do not fit within the philosophical and moral framework about biomarkers that I aim to construct. As should be apparent from the preceding section, genetic biomarkers, especially when many genetic biomarkers are combined into polygenic risk scores or multiplexed with other biomarkers, share many challenges with non-genetic biomarkers. My focus on non-genetic biomarkers is partly practical: to draw attention to this under-discussed source of risk information. In the wake of the sequencing of the human genome, most of the discussion on the ethics of predictive biomarkers has been restricted to genetic biomarkers, while discussion of non-genetic biomarkers in psychiatric and neurological disorder is relatively new.

I adopt this focus also because legal and political structures carve out genetic biomarkers from non-genetic ones. There are policies in many countries that prohibit certain uses specifically of *genetic biomarkers*, for example, the Genetic Information Non-discrimination Act (GINA) in the United States. I am skeptical as to whether there are differences between genetic and non-genetic biomarkers that are morally significant enough to make such policy distinctions. I do think that there *can* be morally relevant differences between genetic and non-genetic biomarkers, but they are not of the sort traditionally envisioned. For example, I highlight in Chapter 6 that predictive biomarkers can be morally preferable if they identify short time windows of elevated risk of a harm, rather than long time windows; the shorter the time window of risk, the less burdensome it may be for the individual to take precautionary measures against the risk manifesting in negative consequences to others (consider avoiding driving if one has an elevated risk of seizure). Because genetic markers are present from fertilization, they are more likely to be associated with long time windows (e.g., whole lives) than non-genetic biomarkers, which may emerge late in life. Of course, this is not always the case, as some genetic biomarkers can increase the risk of harm with a very characteristic age of onset, and some non-genetic biomarkers like high blood pressure can also have very long time windows. Though I will not investigate in depth the potential differences between genetic and non-genetic biomarkers, I have included a very short and preliminary discussion in Appendix I.

MENTAL RETARDATION AND PHENYLKETONURIA: A HISTORICAL ILLUSTRATION OF TYPES AND PROPERTIES OF PREDICTIVE BIOMARKERS

The following example is intended to consolidate the background on types and properties of predictive biomarkers introduced previously. A secondary function of this example is to illustrate the heterogeneity of many existing categories of disorder and the role of history.

On the path toward the development of treatments, we will often first discover a biomarker, and only after further rigorous study will we discover whether and under what conditions manipulation of that biomarker is therapeutic. Each year, approximately 300 babies are born in the United States with a metabolic dysfunction called phenylketonuria (PKU) (Centerwall & Centerwall 2000). These infants accrue elevated levels of the amino acid phenylalanine in their blood; if the dysfunction is not identified and treated, most show developmental delay within months, and are moderately to severely mentally retarded by their fourth year. The severity and exact symptoms vary significantly across this set of children, however, with some suffering irritability or convulsions, and some exhibiting "destructive behavior" (Centerwall & Centerwall 2000). If untreated, some die quite young, while others live into their twenties.

Applying the biomarker terminology from the previous sections, we can identify elevated blood phenylalanine levels as a predictive biomarker of mental retardation. It is a near perfect predictor, as very high levels almost always lead to some mental retardation; however, if we change the prediction from simply "some mental retardation" to mental retardation with convulsions, or mental retardation with destructive behavior, or mental retardation with convulsions and destructive behavior, we could make blood phenylalanine a very imperfect biomarker (decrease its predictive value). By doing so, we can see that the complexity of what is predicted has a large impact on the predictive value of a biomarker. We will return to this point shortly, to discuss the heterogeneity of psychiatric definitions in the American Psychiatric Association's *Diagnostic and Statistical Manual of Mental Disorders* (*DSM*).

Subsequent research into the genetics of PKU has found that those who suffer from it have one of a wide variety of rare genetic variations (most, but not all, in the human phenylalanine hydroxylase gene) that lead to the inability to metabolize phenylalanine (Scriver & Waters 1999). This inability leads to a toxic buildup of phenylalanine and progressive and irreversible brain damage. With this evidence, we can redefine high blood phenylalanine as a *causal* predictive biomarker of severe mental retardation. However, with so many different and rare mutations causing the same functional deficit (inability to metabolize phenylalanine), researchers looking for genetic mutations rather than metabolites might have completely missed the association, or else required an impractically large sample size to detect them. This illustrates how a higher-level biomarker (blood phenylalanine) is able to summate the effects of many different lower-level biomarkers (genetic variants).

Phenylalanine makes up about 5 percent of the protein in a Western diet; a particularly good source of phenylalanine is milk. Phenylalanine enters the body in protein foods and then the children with PKU cannot use or eliminate it efficiently. As early as the mid-1950s, doctors discovered that if these children were placed on a special diet low in phenylalanine, mental retardation could be avoided—but this was true *only if the children were started on the diet in very early infancy* (Centerwall & Centerwall 2000). Started too late, the diet fails to preserve patients' mental capacity. Blood phenylalanine levels can be reduced later in life, but by then the child has already undergone significant brain damage. Lowering the level of blood phenylalanine is most therapeutic immediately after birth and decreases in therapeutic value steeply thereafter.

The case of PKU also highlights that when people call a genetic or other biological variation "deterministic" in regard to a disorder, they mean that the variation is deterministic in the range of environments typically encountered, or the ones researchers have studied. PKU, for example, is determinatively related to mental retardation in an environment that includes an average diet, but not a diet low in phenylalanine. Even untreated, however, some patients do not exhibit cognitive delay (Scriver & Waters 1999).

The Norwegian chemist and doctor Asbjørn Følling first discovered the metabolic dysfunction in PKU not through a blood biomarker but through a biomarker in urine (Centerwall & Centerwall 2000). When phenylalanine reaches high levels in the blood, "like water over a dam, some of it spills into the urine as phenylpyruvic acid and other chemicals" (Centerwall & Centerwall 2000). When the normal metabolic pathways of phenylalanine fail, it is processed down other pathways, some of which are toxic, and one of which is secreted in the urine as phenylpyruvic acid. Phenylpyruvic acid turns green when combined with a few drops of 10 percent ferric chloride solution; this is how it was first detected.

High urine concentrations of phenylpyruvic acid, therefore, are also a predictive biomarker of mental retardation (and an imperfect predictive biomarker of mental retardation with destructive behavior), but not a causative biomarker. Yes, the low phenylalanine diet lowers blood phenylalanine and eliminates phenylpyruvic acid from the urine, but one could imagine a treatment that eliminated phenylpyruvic acid from the urine without treating or preventing the correlative mental retardation, for example, impairing the transport of metabolites into the urine (which would drive blood levels up faster) or causing excessive production of urine and thus a dilution of the biomarker to undetectable levels.

The circumstances of PKU's discovery illustrate the significant role that disorder classification plays in the discovery of biomarkers. If the mechanism of PKU were unknown and a researcher interested in the molecular basis of infant mental retardation were to run a mass proteomic analysis study, she would at best identify elevated blood levels of phenylalanine as a biomarker of small effect across the whole population of mentally retarded children; infants who become mentally retarded would be slightly more likely to have elevated levels of blood phenylalanine than other children. When Dr. Følling made his discovery, mentally retarded children all across the world were tested with the ferric chloride solution and only 1–2 percent exhibited the elevated levels of phenylpyruvic acid (Centerwall & Centerwall 2000). If a sample population contains heterogeneous etiologies, as does the set of people with mental retardation (or maybe schizophrenia, bipolar disorder, or depression), even a causal predictive biomarker with nearly 100 percent penetrance[4] may be identified as a weak association or may be missed entirely if it is a rare variation. Remember that of the 1–2 percent of mentally retarded children with elevated urine phelypryruvic acid, only a sliver would share the same causative *genetic* variation (Scriver & Waters 1999). The most sensitive genome-wide association screens would be unable to detect such rare causal variants (Gibson 2012). Multiplexing can partially address this limitation in sensitivity, but our confidence in the moral weight of a classical diagnosis should be tempered by remembering that current categories of disorder and known biomarkers exist with a large pinch of historical chance (e.g., due partially to the idiosyncratic strengths and limitations of the clinical tests and biological assays that have historically been available).

With the key characteristics of bioprediction introduced, I will narrow the scope of this book to the bioprediction of brain disorder.

BIOPREDICTION OF BRAIN DISORDER: SCOPE AND SPECIAL ISSUES

Bioprediction may be applied throughout the medical sciences and beyond (for example, the bioprediction of voting behavior). While the issues discussed in this book may also be relevant to these other types of bioprediction, I intend to address the bioprediction of brain disorder. This limit is partly practical, partly based on my own scientific and philosophical interest in brain disorders, and partly because I view ethical and conceptual complications of bioprediction as most likely to arise in this area (see "What's Special About Bioprediction of Brain Disorder?" later in this chapter).

I use "brain disorder" as a broad term encompassing most psychiatric and neurological disorders. I include both psychiatric and neurological disorders because their symptomatic presentation is often indistinguishable; anti-NMDA receptor encephalitis, for example, often presents with a set of symptoms indistinguishable from psychosis (Barry et al. 2011; Dalmau et al. 2008; Dalmau et al. 2011). Furthermore, as biomarkers increasingly enter psychiatry, the boundaries between psychiatry and neurology are likely to erode. The definition of disorder will be the topic of my next chapter, and so I will say no more here.

Bioprediction of brain disorder, as discussed in this book, excludes the prediction of traditionally somatic events, like ovarian cancer or blocked blood vessels in the heart (heart attack), but might include related biomarkers insofar as they could provide probabilistic information on the likelihood of lost or altered consciousness, for example, related to a blocked blood vessel (stroke) in the brain. Moreover, a somatic condition, like certain types of ovarian teratomas, might itself serve as a biomarker of brain disorder; in Rasmussen encephalitis, a certain type of ovarian teratoma produces antibodies to a brain protein and increases the risk of seizure (Nestler et al. 2009); in anti-NMDA receptor encephalitis, the teratoma predicts psychosis symptoms (Dalmau et al. 2008).

For reasons that will become clear, I do not restrict the bioprediction of brain disorder to bioprediction of brain disorders *as currently categorized*; I include bioprediction of harmful components of existing disorders and of sets of harms that cut across the boundaries of existing disorders. Since seizure can be seen in a variety of conditions besides epilepsy, including PKU, fragile X syndrome, schizophrenia, and others, I will consider the bioprediction of seizure even if the biomarker has zero specificity to a particular named disorder.[5]

I will exclude prediction based on environment alone (as this does not involve the use of predictive biomarkers and as such is not *bio*-prediction), although I will include predictions based on an interaction of a certain environment with a certain biomarker (i.e., strobe lights interacting with a biologically defined variation to increase the likelihood of seizure, or a biomarker of amphetamine-induced psychosis). Though voting and other social behavior may become at some point inquiries for bioprediction, I restrict myself to bioprediction of behavior that is harmful to the individual; the role and definition of harm will be discussed in Part I, Chapter 4.

I do discuss bioprediction of certain extreme forms of "bad" behavior, namely, impulsive aggressive behavior, but I do so with a focus on the increased risk of harm it can cause to the person exhibiting that behavior (e.g., risk of incarceration). While I do not discuss the bioprediction of things that would fall under "superior brain performance," such as high IQ, sometimes bioprediction that defines a group as high risk for a harm also defines low-risk groups.

The philosophical basis for these exclusions will become clear once I have developed my concept of disorder, for the excluded phenomena will not count as disorders.

What's Special About Bioprediction of Brain Disorder?

As with the distinction between biomarkers and other sources of risk information, the differences between brain disorders and other disorders are best thought of as differences in *degree* rather than *kind*. I highlight here three qualities that are more salient in brain disorders: (1) the degree to which diagnoses are subjective, symptomatic, and all-or-nothing, (2) the degree of perceived association with voluntary behavior and morality, and (3) the degree to which there is a history of discrimination.

Diagnoses in Psychiatry

The current state of diagnosis in psychiatry is based largely on subjective semi-systematic interviews and third-person reports about a patient's behavior, emotions, or beliefs. Loose clusters of symptoms are tethered to diagnoses as defined by the fifth edition of the *Diagnostic and Statistical Manual of Mental Disorders* (*DSM-5*) or the *International Classification of Disease 10* (*ICD-10*). To qualify for diagnosis with a disorder, a person needs to exhibit a defined number of symptoms on the category's list. Each category contains substantial heterogeneity; patients with the same diagnosis often experience very different symptoms. Similar symptoms, moreover, appear across multiple categories. Despite this heterogeneity and overlap, the disorders are often by definition mutually exclusive and binary: either schizophrenia or bipolar disorder, either mentally disordered or healthy.

For our purposes, it is also significant that the diagnoses make no reference to any underlying biological alteration. Because of this, the introduction of biomarkers within psychiatry faces a steeper uphill battle than in other medical areas. Existing diagnostic categories for psychiatric disorders, for example, may be less likely to map onto discrete sets of biomarkers than disorders, say, in cardiology, which already have surer footing in biology and consequently risk being discarded as "nonspecific" even if they provide information on a risk that cuts across existing disorders.

A second factor that contributes to the grade of the uphill slope is the cultural emphasis of psychiatric disorders as conceptually different from "physical disorders" even if (as I think to be the case) there are few actual differences. As one group describes, for example,

> Psychiatric conditions also come with the special challenge that they have traditionally been considered to be disorders of the mind. Thus, convincing clinicians and the regulatory agencies that blood-based assays are sensible and can be predictive is most likely the biggest challenge of all. (Bahn et al. 2011)

In the not-so-distant past, moreover, psychiatric disorders had "organic exclusion clauses," that is, if someone could point to a likely "organic" (i.e., biological) cause, then the set of symptoms was deemed to be not a mental disorder; the patient would consequently be deported to neurology. Though the organic exclusion clauses have now been rightly excluded from psychiatric diagnostic criteria, one need not look far to find vestiges of this way of thinking, one example being the carve-out of mental-health coverage from large health insurers in the United States until the 2008 ruling that required parity in coverage.

Diagnoses within neurology are relatively free of these concerns, as the field uses both the appearance of behavioral, emotional, or cognitive symptoms and macro-level brain abnormalities, such as lesions (areas of dead or dying tissue) detected with brain imaging. Neurology, nonetheless, faces the challenge of having an abundance of retrospective biomarkers, and very few predictive ones.

Morality Is More Salient with Brain Disorders

While many medical conditions raise issues related to moral responsibility, these issues are most salient in the realm of brain disorders because of their more direct relationship with behavior, choice, and personality (Radden 2002). For example, brain disorders can change the likelihood of actions to which we normally ascribe intent, blurring the lines about how an individual should evaluate his own actions and how others should respond to those actions. To take a neurological example, consider that in Huntington's disease, deterioration of the caudate nucleus can increase the likelihood of behavioral disinhibition; in a hypothetical case in which a woman with early Huntington's disease shouts bad language at her adolescent son who had accidently spilled a glass of water on her, how should the boy respond? Is it appropriate for him to feel hurt, or to shout back at his mother, or to ask for an apology? Again, this blurring is not *unique* to brain disorders—many cancers (or their treatments) are painful, and being in pain can similarly increase the likelihood that the person will have an angry outburst—but if it is accepted that this blurring happens more often or to a higher degree in brain disorders, then bioprediction in this area should inherit enhanced blurring.

A History of Discrimination, Maltreatment, and Blame and a Present of Unmet Need

The history of the treatment of the mentally ill, at least in Western societies, is particularly terrible. People were jailed, beaten, persecuted, or at best ostracized for their strange behaviors. While the mental hygiene movement, started by Clifford Beers in 1908 with the publication of his book *A Mind That Found Itself* (Beers [1908]1981), did much to stop the physical restraint of patients as a common treatment and generally improved the conditions of institutions, the negative images of involuntary commitments and institutional abuse remain common in the media (for example, *One Flew Over the Cuckoo's Nest*,

Table 2.2 BIOPREDICTION: GLOSSARY OF TERMS (IN ORDER OF APPEARANCE)

Term	Definition	Example
Bioprediction	The use of a biomarker to predict the likelihood of something occurring in the future i.e the use of a predictive biomarker	Measurement of blood pressure (biomarker) to estimate 10-year likelihood of heart attack or stroke
Biomarker	A (biological) characteristic that is objectively measured and evaluated as an indicator of normal biological processes, pathogenic processes, or pharmacologic responses to a therapeutic intervention	Blood pressure, plaques in the brain, size of cancerous growth, number of sites of cancer metastasis, cholesterol level, quantity of an inflammatory marker
Predictive biomarker	A biomarker used to estimate the likelihood of a particular future event	Abnormal heart rhythm to estimate the likelihood of cardiac arrest and death
Real-time biomarker	A biomarker used to estimate the likelihood that a patient is currently experiencing a particular event	Abnormal heart rhythm to estimate the likelihood that there is compression of the rhythm-generating nerve fibers in the heart
Retrospective biomarker	A biomarker used to estimate the likelihood that a patient has already experienced a particular event	At autopsy, observation of a growth of cells next to the bundle of rhythm-generating nerve fibers in the heart, used to hypothesize that compression of those nerves likely led to arrhythmia, cardiac arrest, and death
Correlative biomarker	A biomarker with an unknown relationship to the causal chain of the event of interest	All biomarkers when making judgments at an individual level or specific event
Causative biomarker	A biomarker that is at least a partial cause of the event of interest	High blood pressure and heart attack in a particular population
Non-causative biomarker	A biomarker that is not involved in causing the event of interest	Ache in my bones predicts, but does not cause, rain
Surrogate Endpoint	An endpoint, other than survival or irreversible morbidity, that is reasonably likely, based on epidemiologic, therapeutic, pathophysiologic, or other evidence, to predict clinical benefit.	Blood pressure lowering as a surrogate endpoint for morbidity and mortality associated with heart attack

or Rhiana's music video, "Disturbia"). Partially because of our historically poor understanding of mental illness, there is a high degree of stigmatization; as people become more familiar with the illnesses, this stigma hopefully will decrease (Radden 2002). Stigmatization and a history of discrimination, of course (and unfortunately), are not unique to mental disorder—they are seen in many types of illnesses—but could reasonably be argued to have tinted mental disorder *particularly strongly*, and for this reason we should approach the bioprediction of brain disorder with an eye more sensitive to stigmatization than we would with other disorders. This history is also part of why I argue for the explicit inclusion of justice in the definition of disorder Chapter 4.

Finally, the current unmet clinical need of those suffering from mental illness is staggering. Unipolar depression, schizophrenia, and bipolar disorder together account for 96.7 million disability adjusted life years (DALYs) globally (Collins et al. 2011) and mental disorders (including substance abuse disorders) are the leading cause of years lost to disability (YLD), accounting for 22.9 percent globally (Whiteford et al. 2013). Mental disorders are disabling not only to individuals but also to nations, as they represent the leading loss of economic output of all non-communicable diseases (Bloom et al. 2011). Though the Affordable Care Act (ACA) will drastically increase individual access to mental health treatment, the blunt medications to which we can offer access work only for some patients, for a subset of their symptoms, and often work only for a time, leaving far too many individuals on an accelerating cycle of hospitals and failed drugs. Despite the incredible need for better treatments, many pharmaceutical companies have recently abandoned their psychopharmacology research and development programs, citing the lack of basic mechanistic understanding necessary to create effective therapeutic hypotheses (see Hyman 2012). These circumstances combine to make those suffering from mental disorders especially vulnerable.

This cultural context of brain disorders is important to keep in mind as we continue in this discussion about the neuroethics of biomarkers. Next, in Part I, I will show how we should reorient our concept of disorder around probabilities. Such a conceptual reorganization is integral to approaching with clarity the ethical implications of bioprediction of brain disorder.

NOTES

1. Prognostication based on a biomarker or set of biomarkers is a type of bioprediction.
2. Sometimes, biomarkers are described as "surrogate measures" of pathologies (Ewers et al. 2011). Most of the assays in the traditional practice of the medical specialty of pathology, however, are also surrogate measures of pathologies. A color change in a reactant is a surrogate marker for the presence of blood in fecal matter

(as a surrogate marker for some other process), an increase in sediment in the urine is a surrogate for a change in renal function. Even the plaques and tangles of Alzheimer's fame are surrogate markers for brain pathology in the form of a staining pattern visualized by a silver stain.

3. This quotation may also reflect a tendency to refer to something as a biomarker only when one does not yet have information on whether it is causally related to what it is marking. However, biomarkers can also be causes.
4. Penetrance is a measure of how often a phenotype occurs together with a marker versus how often the marker is seen without the phenotype.
5. This philosophy, in some ways, is similar to a "problems-based approach" of stratification (Filiou & Turck 2011).

PART I

Reorientation of the Concept of Disorder

In the early stages of the writing of this book, I began by considering individual ethical issues raised by bioprediction, but I slowly realized that I could not move the discussions forward without first addressing the implications of bioprediction for theories of disorder. I found myself responding to similar debates over and over again, a phenomenon that I ultimately concluded reflected unspoken assumptions about the higher-level ideas of what we should or should not call disorders and how we should categorize them. Culver, Gert, and Clouser state succinctly the relationship of the theory of disorder with the ethical sphere: "the concept of [disorder] that one adopts has wide ranging repercussions for everything from establishing the goals of medicine to the distribution of health care in society" (Gert et al. 2006, 135). It is on assumptions about the concept of disorder that the specific ethical arguments turn. It is consequently important that we closely examine these assumptions, and the evidence for and against making them.

Though disorder is often thought of or treated as entirely present or absent (binary), I argue that the development of biomarkers makes it clear that we should reorient the concept of disorder around probability. I call this a reorientation because probability has always existed at the core of the concept of disorder; the development of biomarkers provides the opportunity to make this connection more explicit. This centrality is especially clear around disorders that are periodic in nature (epilepsy predicts risk of seizures) and ones that have little present harm but significant risk of serious adverse event (hypertrophic cardiomyopathy predicts risk of sudden cardiac death). If we accept this thesis, existing disorder categories are best thought of as a sort of "folk bioprediction," where instead of biomarkers, more subjective signs and symptoms have served as imperfect indexes of (a) the likelihood of future harm, (b) the relation of that harm prediction with time (e.g., whether it will probably last for a day or a lifetime), and (c) the likely identity of that harm. Thus predictions of future harms based on bioprediction are relevantly similar to the predictions that follow from diagnosis, except that prediction with biomarkers will often be made actuarially rather than clinically. It should in many cases make little difference whether our predictions are made with a set of biomarkers, a set of signs and symptoms, or a combination.

What *do* matter are the various qualities of our predictions. For example, in this reorientation it becomes clear that we should be very specific about the magnitudes of the risks, their temporal characteristics, and when we are discussing probability of harms somewhat directly (e.g., risk of death) or indirectly (as would be the case if a set of biomarkers predicts a diagnosis that we care about because it signals a higher likelihood of experiencing a different set of harms). We may be able to best identify and discuss these salient features if we integrate all the available sources of information, biomarkers and otherwise, into quantitative representations of probabilities.

Existing named categories inform us simultaneously about many different likelihoods for many different specific harms, and thus their relevant features are sometimes obscured. If we find, therefore, that we justify the existence or the demand of certain moral obligations by a *diagnosis*, we should think carefully about how that diagnosis is related to *types of risk*, which feature of the risk is doing the work in our justification, and given which risk is doing the work, whether we should amend our attitudes toward situations that carry similar risks but are estimated through different means, for example bioprediction. If we are to stick with a naming system of discrete disorders, therefore, we should strive to modify our naming in a way that most accurately captures the salient features of risk, and we should recognize the names as heuristics for risk estimation.

Part I, which includes Chapters 3–5, addresses the challenges that the development of predictive biomarkers raises for the concept of brain disorder. In Chapter 3, I use a fable and the story of Shakespeare's *King Lear* as analytic lancets to cut into our current concepts of individual disorders within psychiatry. I argue that the specific disorders, established first as useful heuristics, have drifted away from what they should be useful *for* and have become reified. This happens because we tend to operate in the areas where our tool (current categories of disorder) works best (in a binary or non-probabilistic way), rather than where we are likely to be most successful at the enterprise we started out on (helping people). The development of biomarkers provides an opportunity to recognize this reification, and to move away from thinking of disorders as individual silos by reorienting our general concept to handle the explicit probability of harms. In the past several years, the idea that many disorders, depression among them, might be better thought of in dimensional terms of the sort currently used for diabetes and hypertension has gained traction (Hyman 2011; Karlawish 2010; Vickers et al. 2008). While this shift is in the right direction, it is not a fundamental enough change in thinking to help us parse the intersections of bioprediction, responsibility, and distributive justice.

In Chapter 4, I move toward such a fundamental conceptual reorientation of disorder by engaging directly the three major competing theories within the philosophy of medicine. I argue that in light of the development of biomarkers, each theory is inadequate. I propose a reorientation of the concept of disorder around the probability of harm. I defend my reoriented concept, and show that rather than being threatened and confused by the development of bioprediction, it is fortified. This reorientation realigns the concept of disorder with the goal of medicine to reduce human suffering. Through this analysis I raise concern about the proposal (which was nearly adopted) of the *DSM-5* Task Force to remove the clause about risk of harm from the concept of mental disorder in the *DSM-5*.

In Chapter 5, I illustrate some of the benefits of conceiving of disorder in terms of probability of harm, one of which is the possibility to use specific disorders as heuristics to conceptualize "risk bands," that is, levels and types of risk of harm. I show that this reorientation is sufficient to address many of the practical concerns in debates about how to handle high-risk states for first psychosis and dementia.

As I wrote in the guidance to readers ("The Organization of This Book") at the end of Chapter 1, I have structured the work so that readers should be able to understand the discussion of the practical issues of Chapters 5–8 with only the thin version of a risk-centered concept of disorder that I have by this point in the book highlighted. For those readers who are interested in a deeper dive into the theory of disorder, Chapters 3 and 4 include a detailed explication of my justifications for this risk reorientation. By including this philosophical analysis before addressing the specific ethical implications of bioprediction, I hope to construct an overarching clarity that we can carry forward in the consideration of the implication of bioprediction for moral responsibility and distributive justice.

3

"There Is More Light Here"

Re-illuminating the Categories of Mental Illness

LUX SANS VERITAS: NASRUDIN[1]

> Someone saw Nasrudin scouring the ground.
> "What have you lost, Nasrudin?"
> "My key."
> After much fruitless searching, the helper asked: "Where did you drop it?"
> "At home."
> "Then why are you looking here?"
> "There is more light here."
>
> *(adapted from Shah 1966)*

What can we learn from fabled Nasrudin? He surely seems foolish for not looking for his key where he dropped it, at home. Yet there is reasonableness to his mistake, as we shall see. And, as the development of biomarkers helps us to realize, perhaps we are committing a similar mistake in our thinking about mental disorder. I will use Nasrudin and one other example to illustrate this possibility by analogy and then will make the connection with the concept of disorder explicit.

To begin, let us consider how Nasrudin *could have* come to arrive at this point, in the street looking for a key that he dropped at home.

Imagine Nasrudin's home: a tired place where an oil lamp struggles against darkness. Suddenly, Nasrudin realized he had lost his key. He reasoned that it must be at home, as he used the key to enter. But where? Perhaps it fell as he reclined in his sagging chair. And so, Nasrudin fumbled a time there. As little light penetrates this corner, however, Nasrudin was forced to search by feeling. This method was difficult, reasoned Nasrudin; it would be much easier to search by seeing. So Nasrudin moved toward the sputtering oil lamp, where he did indeed find searching much easier.

Soon, however, the dim light strained his eyes. Then, Nasrudin noticed a glow near the door. Moonlight! And so from lamplight to moonlight to greater moonlight and from moonlight to streetlight went Nasrudin, always nose to ground, always thinking how each new spot was surely better for the activity of searching. And surely, bright ground beneath a streetlight *is* a better place for looking.

Enter: Nasrudin's helper.

What Nasrudin forgot is that a good place-for-looking need not coincide with a likely place to find that for which one looks. Initially, this did not matter, as the good place-for-searching and the good place-for-searching-for-his-key were closely correlated. At some point, however, they drifted apart. But Nasrudin was already blinded by his heuristic: that he should look in the place with more light. He did not notice the drift.

SYMBOLS WITHOUT SUBSTANCE

What is more difficult than descrying another's mind? Yet we try every day (especially within psychiatry). Little is more important. We are forced in this endeavor, like Nasrudin, to rely on correlates. We rely on words and acts to guess the thoughts and emotions that really matter. Keeping the nose too close to these correlates, however, opens us to the risk of ending up like Nasrudin.

This is what happened to King Lear (Shakespeare [1608]1997). At the beginning of Shakespeare's play *King Lear*, the ancient Lear holds a gathering to publicly divide his kingdom among his daughters. Before the division, he requests from each of his three daughters, Goneril, Regan, and Cordelia, a flowery show of words demonstrating the extent of their love for him. Goneril and Regan, who love him not, exhale grandiose avowals of their love that more than satisfy Lear. Cordelia, who loves him, refuses such a public show, an action that leads to Lear's decision to banish Cordelia. Lear, initially convinced of Regan's and Goneril's love, resigns himself to their care, but slowly begins to realize by their actions that they love him not, and that he banished a loving daughter. Such realization drives Lear mad, and his former kingdom tumbles to chaos. Cordelia returns toward the end of the play, at which point she has a reconciliation with Lear. A major theme of the play—and one that is important to the concept of disorder—is Lear's reification of words and symbols, his forgetting that they are mere correlates of the substance of value.

At the beginning of the play, we discover Lear at a moment similar to the one when we discovered Nasrudin, that is, when a logical heuristic has already drifted oceans from what really matters. Nasrudin's mistake becomes clear when he responds to the helper's questions. Lear's becomes visible during his love contest among his daughters to determine the division of his kingdom: "[. . .] which of you shall we say doth love us most, / that we our largest bounty may extend . . . ?" (*Lear*, 1.1.53–54). Goneril and Regan's vocalizations, lengthy and extravagant, satisfy Lear magnificently. Though loveless words, Lear cannot see them as such. Rather, Lear walks into streetlight with Nasrudin. When he asks Cordelia, her failure to produce obsequious speech triggers the unfatherly rage that tumbles them to chaos.

Nasrudin forgot that light is valuable *only insofar as* it helps find his key. Lear has forgotten that words and externalization are valuable *only insofar as* they give voice to interior love or substance. Unique among major Shakespearean characters, Lear has *not one* aside. Since asides are how Shakespeare shares the internal thoughts and struggles of his characters, this absence suggests that Lear has lost sight entirely of the interior. He mistakes the physical manifestation of the inner self for the actual self, thereby reifying words and drifting *ad absurdum* to greater levels of external manifestation and further away from the internal.

The depth to which Lear mistakes symbol for substance becomes clearer when we examine his wish to relinquish all, but retain "the name and all th'addition to a king" (*Lear*, 1.1.137). Just as he presumed value for flowery words, so does he for the kingly name. But lovely words without love and powerful names without power are as light without possibility of key finding.

The character of the fool tries to educate Lear on his mistake, "thou art an 0 without a figure" (*Lear*, 1.4.184). Noughts, 0s, thrust their substance outside and leave nothing in the center, just like Lear. A lone nought is nothing. Infinite noughts are nothing. Yet, with a "figure," a number other than 0, the nought has great value. Add a nought and one becomes 10. Just like the nought, an external manifestation adds value to an internal emotion (the figure) by enabling its imperfect communication.

Similarly, light, by facilitating the search, adds value to an area that already has a probability of key finding by further increasing that probability. But light has no value *alone*. In his long life, Lear has forgotten that it is the figure that matters, not the zero. As the fool suggests, Lear has come to assume that the zero is itself valuable. Lear takes the symbol to be of singular importance and ignores the substance.

In psychiatry, the analogue is to let the name of the diagnostic category obscure the relation of that diagnosis to the harm the patient experiences, and is likely to experience in the future.

DIAGNOSES WITHOUT UTILITY?

Nasrudin's and Lear's follies spill into mental disorder. As explained earlier, doctors currently diagnose mental disorder based on behavioral and emotional symptoms. The historical reason is a practical one: as we saw through Lear's difficulties interpreting his daughters' intentions, it is difficult to peer into another's mind. For psychiatrists, external manifestations have provided the only accessible way of grouping and treating mental disorders (through psychoanalysis or behavioral therapy, for example). Before the advent of less invasive methods to investigate tissues and cells, the same was true for other medical specialties.

The stresses placed on such diagnostic categories by the historical development of medical technology can be illustrated with the following imaginary case: imagine we are interested in erratic car-driving behavior but are unable to examine cars. We might establish two categories: general (erratic all the time), and night-limited (erratic only at night). The latter can be treated by avoiding night driving, and both can be mildly ameliorated by teaching defensive driving techniques. Note that we care about erratic car-driving behavior only insofar as it is related to an increased likelihood of harm to the driver and to others.

Imagine we serendipitously discover that some (not all) general-erratic drivers respond miraculously to an unusual treatment: when we wash the car regularly, the erratic driving disappears! We also discover that this washing treatment works for a subset of the night-limited-erratic drivers! However, we keep our general-erratic and night-limited categories. As we shall see, these categories might resemble schizophrenia and bipolar disorder.

Further physical treatments for the erratic driving are discovered; they work (better than behavioral treatments alone) for only a portion of what we previously thought to be a homogeneous group, or seem to cross boundaries. Meanwhile, we develop techniques to visualize the car. We realize that the windshields of the washing-responsive subgroup attract heavy grime that obscures vision; the night-limited subgroup develops grime that is translucent during daylight, but produces a blinding glare in night headlights. Washing in both cases clears the windshield. Continued adherence to the original groupings increasingly resembles Nasrudin and Lear. The heuristic (behavioral grouping) has drifted apart from that which gives it practical value (managing risk of harm).

Back in reality, a similar development looms with the century-old specter of German psychiatrist Emil Kraepelin. Kraepelin, for pragmatic reasons, defined a rapidly progressive psychosis and a cyclical one, categories from which descended schizophrenia and bipolar disorder (Fischer & Carpenter 2009).

But with the serendipitous discovery of drug treatments (like lithium for bipolar disorder), and the development of tools to better examine biomarkers of these illnesses, it is not clear that these categories should remain. For a portion of those diagnosed with bipolar disorder, for example, lithium is a boon. For others, it works briefly or not at all (Nestler et al. 2009), and they cycle painfully through medications and hospitals. Since lithium works on a molecular level, there must be something different in the molecular biology between patients who respond and those who do not; put otherwise, the diagnostic category is far from homogeneous. The heterogeneity of existing psychiatric categories is further supported by recent work that is elucidating a multidimensional genetic landscape; for example, variations in 108 separate genetic loci (i.e., 108 regions of the genome) were recently implicated in risk of schizophrenia (Schizophrenia Working Group of the Psychiatric Genomics Consortium, 2014).

Meanwhile, new research suggests that subpopulations within what was once considered disparate categories of disorder may have common ground. A Swedish population-register study of 9,009,202 individuals looked at heritance between (instead of within) bipolar disorder and schizophrenia. Surprisingly, first-degree relatives of individuals with one disorder have a significant risk of *either* disorder, rather than the *same* disorder (Lichtenstein et al. 2009). Adopted children whose biological parents had bipolar disorder had *almost identical risk of schizophrenia* as they did of bipolar disorder.[2] This study provides evidence for a portion of genetic susceptibility shared between the diagnoses.

An international collaboration of scientists found further evidence for molecular pathways shared between the silos of mental disorder (The Network and Pathway Analysis Subgroup of the Psychiatric Genomics Consortium, 2015). This work suggested that extending across the categories of schizophrenia, bipolar disorder, and major depression is a shared variation in genes whose products are important in histone methylation (involved in modulating accessibility of DNA regions), immune signaling, neuronal signaling, and postsynaptic elements of the synapse. These shared risk biomarkers and the variations in the genetic elements in this second study are examples of genetic biomarkers that cut across existing categorical silos of disease, reflecting a concept of pleiotropy that will be returned to in Chapter 4.

If we hope to treat these illnesses successfully, we should strive to base our diagnostic categories upon variation at the same molecular level at which our interventions work. To do so requires recognition of subdivisions within, and relationships between, previously reified diagnostic categories. Once we gain the tools to visualize the car, we should not remain blindly committed to earlier definitions of disorders.

Traditionally, the molecular level has been locked inside heads more tightly than Cordelia's silent love of Lear. But new medical research and technology are increasingly cracking the silence. On the striking edge, for example, a new noninvasive technique may give a strong voice to patients' own molecular pathways. A few skin cells can be taken and driven in reverse down the developmental pathway back into undifferentiated stem cells (called induced pluripotent stem cells, or iPSCs), and then caused to differentiate into neurons. Since these cells come from the patient's own tissue, they remain genetically identical to the patient's own brain cells (Dolmetsch & Geschwind 2011). Such a technique may revolutionize diagnostic categories in psychiatry in the same way that safely taking a biopsy of a tissue lump and examining the cellular structure did in oncology.

In 2011, a group of researchers was the first to apply this technique to the study of a complex psychiatric disorder. These researchers showed that skin cells transformed to neurons from patients diagnosed with schizophrenia showed a characteristically abnormal connectivity (Brennand et al. 2011). Interestingly, treating the cells with a specific antipsychotic medication rescued normal connectivity.

If validated, such technology could, in theory, be automated and used to screen cells from multitudes of patients; cycling between laboratory and clinic, we might thereby tease apart patient subgroups according to characteristic disruptions of cellular pathways (Buxbaum & Sklar 2011). In a not-so-distant future clinic, a patient's own cells might aid in a more personal diagnosis and medicine. The characteristic disruptions, which are biomarkers, could potentially become robust predictive biomarkers of the onset or reappearance of harmful symptoms if the person continues in her current environment; the biomarkers could also help to estimate the likelihood that those risks-of-harm would be reduced if the persons molecular environment were to be altered, for example, by the presence of a molecular medicine.

WE ARE NASRUDIN'S HELPER

What really matters is what Nasrudin's helper does once she realizes that "more light" is useful only in the context of "where the key might be," and "where the key might be" is not here in the street.

We have been following the previously established diagnostic categories because collections of *present* symptoms were at one time our best practical way for making categories that enabled the medical profession to help people, because these categories are where we found Nasrudin searching. But, in light of current research, there is a responsibility to question whether this sort of categorization is still useful or should be discarded, especially as multiplexed biomarkers may enable groupings that are based explicitly on risk in a way that does not map onto these existing categories.

In this re-evaluation, we should learn from Cordelia, who does speak up:

> Unhappy that I am, I cannot heave
> My heart into my mouth.
>
> *Lear*, 1.1.91–92

Rather than spew forth flowery words, she rejects them. But this is spiteful rejection of *all externalization*, an angry assertion that "my love's / more ponderous than my tongue" (*Lear*, 1.1.77–78).

There is a risk of straying too far to the other extreme, of telling Nasrudin to pay *no attention to light*. But shedding more light *in Nasrudin's home* will surely help him find his key. Similarly, careful consideration of present symptoms combined with biomarkers may be better than either alone in capturing the current and future suffering of patients. It is the goal of reducing human suffering, after all, that is most central to the practice of medicine; this practical application is a primary reason to categorize disorders of which we should not lose sight. What is important, I will argue, is that the markers of disorder link strongly to risk of harm (which motivates the concern in the first place), regardless of whether the markers used are those of historically established diagnostic categories or are new bio-actuarial tools. Though the development of bioprediction may enable ways to link with harm in an entirely dimensional way, moreover, there may still be benefits to retaining specific categories as heuristics for magnitude and type of harm.

The question is how to strike the right balance, how to shine the right light in the right place in the right way. An answer will become more pressing as predictive biomarkers become clinically available.[3]

NOTES

1. Nasrudin, an Islamic priest or mullah, appears in a long series of fables popularized by Idries Shah (1966, 1971). Nasrudin does silly, foolish, or otherwise absurd things that illustrate a bit of wisdom. The particular bit of wisdom is usually up for debate; the adapted fable here is useful in thinking analogically about how the concept of disorder, like the principle of looking in areas with more light, may have drifted away from its initial practical purpose.
2. Relative risk 4.3, 95 percent confidence interval 2.0–9.5 for schizophrenia; relative risk 4.5, 95 percent confidence interval 1.8–10.9 for bipolar disorder (Lichtenstein et al. 2009).
3. An early version of this chapter won the Nasrudin essay prize at Balliol College, Oxford, which acquainted me with the incomparable Mullah Nasrudin. My reading of *Lear* was largely developed in Harold Bloom's Shakespeare seminar at Yale.

4

The Probability Dysfunction

In the introduction to Part I, I wrote about the importance of the concept of disorder to the functioning of medicine, moral and legal responsibility, and for determining how to distribute resources. In this chapter I outline my risk-centered concept of disorder and argue that it should be preferred to existing concepts. I need to say something, therefore, regarding my assumptions about how one might reasonably *choose between* concepts of disorder, that is, how one might evaluate what makes one good and another bad and in what context. I humbly submit the following: A concept of disorder should be preferred to the extent it is *useful*.

There are many ways in which to evaluate usefulness; I will briefly describe here several of my assumptions; illustrations of specific features of usefulness will be appear elsewhere in this book. In order to be useful, a concept of disorder must have some element of face validity, that is, it should not be so radically different from existing concepts as to be unrecognizable as a concept *of disorder*; consequently, one measure for usefulness would be how easily the concept can handle conditions that would be described by a lay person as "clear cut cases". A concept of disorder should also have a strong element of construct validity, however, which involves an articulation of what, precisely, makes "clear cut cases" clear cut. Such an articulation will depend on intuitions about what a concept of disorder *should* do. For example, one intuition I write from is that a concept of disorder should allow us to name, with as much precision as possible, the sense in which a particular disorder is 'bad" (and to what degree), that is, *why* a person should avoid, manage, treat it, etc., and should enable a practical evaluation of the degree to which such measures are successful. A second intuition is that a concept of disorder should be useful for guiding action, whether actions of individuals or institutions.

Finally, one practical consequence of a usefulness criterion is that we should strongly prefer concepts of disorder that are robust over context and time. Because usefulness, however defined, is itself context dependent, a concept of disorder should be preferred to the extent it is useful across the domains it is used, for example, in medicine, law, moral responsibility, and priority setting. Because usefulness is also time dependent, a concept of disorder should be preferred to the extent it is *resilient to a future in which each of us has, or is capable of having, a tremendous amount of data from biomarkers and other sources on our likelihoods of experiencing each of a tremendous number of biological events.*

TWO CHALLENGES THAT BIOPREDICTION POSES FOR CONCEPTS OF DISORDER: PLEIOTROPY AND PROBABILITY

Before we can address the implications of bioprediction in moral domains that rely on a certain theory of disorder, we will save time and increase clarity if we first consider the challenges that bioprediction poses for that theory. Chapter 3 contained an extended illustration of the risk of reifying existing disease categories, and suggested reorienting diagnostic categories toward emerging molecular biomarkers. If we hope to find more than a temporary solution, however, we need not only a reorganization of existing categories, but a reorientation of our overarching concept of disorder.

Chapter 3 highlighted that the overarching concept should enable us to deal systematically with biomarker pleiotropy (that a biomarker cuts across existing silos of disorder and even into the silo of what is now thought of as perfect health). Meeting this condition is the first major challenge posed by the development of predictive biomarkers.

"Pleiotropy" traditionally refers to a variation in a single gene that has diverse, often not obviously related, phenotypic effects (Nussbaum et al. 2007, 116). For example, variations in the gene *fibrillin-1* can lead to the appearance of a seemingly unrelated set of symptoms: the great height and body dysmorphology seen in Marfan syndrome (think Abraham Lincoln), increased likelihood of death due to aortic aneurysm, and increased likelihood of vision impairment due to detachment of the retina (Kumar et al. 2010, 144–145). My use of "pleiotropy" expands somewhat on the traditional definition, to incorporate cases in which one predictive biomarker (not necessarily a genetic biomarker) can signal an increased likelihood of multiple events. For example, a biomarker may be associated with an increased likelihood both of harm to self and to periods of great creativity across a population with the biomarker.

Put another way, the challenge of biomarker pleiotropy is the challenge of capturing the emerging interrelatedness of adverse events, current disorders, and biological states. Current concepts of disorder that categorize on the basis of phenomenology, as psychiatry does, incorporate pleiotropy particularly poorly. If disorder A (e.g., schizophrenia) is defined by a set of symptoms, and disorder B (e.g., bipolar disorder) by a set of partially overlapping symptoms, the current framework may say you have disorder A *or* B *or* neither. If a biomarker comes around that is associated with increased likelihood of some of the events in both A and B, the current framework may declare it a nonspecific marker, rather than consider that this is a biomarker with pleiotropy. The conceptual insulation between existing categories discourages thinking about the interrelations that biomarkers suggest. There is no *a priori* justification for believing that biomarkers will map cleanly onto diagnostic categories arrived at by historical accident.

The second major challenge that biomarkers raise for binary concepts of disorder arises when a set of biomarkers blurs the distinction between the silo of an existing disorder category and the meta-silo of non-disorder. I call this the challenge from probability. Existing concepts of disorder emphasize an all-or-nothing binary; you are ill or well, you have "symptoms" or you do not. You have Alzheimer's disease or you do not. You have cancer or you do not. You are a 50-foot oak tree or you are an acorn. The implicit idea is that disorder is a steady state. Biomarkers challenge this idea by revealing the intermediate and uncertain world of 12-inch saplings. The difficulty of dealing with explicit probabilities is already rippling to the surface in the emerging debates surrounding conditions of high risk for the development of psychosis or dementia. These two debates will be addressed in detail in Chapter 5.

The binary model can lend itself to the further reification of existing disorders by encouraging the assumption that the biomarker should be thought of as a diagnostic test for that specific disorder; it correctly identifies some as having the disease early and mis-categorizes some healthy people as disordered. Such a test certainly seems like it could be problematic, if we retain one way of treating those with the disorder and another way for healthy people. As a preview of what is wrong with this thinking, imagine treating everyone with high blood pressure the way we treat those with advanced heart disease. Even thinking in terms of "high" and "low" blood pressure creates a rigid dichotomy, when the risk may be better thought of in continuous terms.

THE NO RISK SOLUTION: THE *DSM-5* WORKING GROUP'S ARGUMENT THAT THERE IS NO PLACE FOR PROBABILITY IN A CONCEPT OF DISORDER

One extreme response to the difficulty of incorporating probability, or risk, into the concept of disorder is illustrated by a proposed (and nearly adopted) revision of the definition of disorder in the *DSM-5*. The response, which I will show to be untenable, is to define away risk by removing what I refer to as the "risk clause" from their definition of mental disorder. I reproduce the key elements of the previous *DSM* definition and the proposed revision here (the risk clause is italicized):

> *DSM-IV-TR*: A disorder is a dysfunction in the individual "associated with present distress (e.g. a painful symptom) or disability (i.e. impairment in one or more important areas of functioning) *or with a significantly increased risk of suffering death, pain, disability, or an important loss of freedom.*" (American Psychiatric Association [APA] 2000)

The *DSM-5* Task Force (which proposed the revision) explains:[1] "The phrase on risk is omitted in order to differentiate more clearly between disorders and risk factors."

Does the medical world exist as a binary of risk factors and disorders? Traditionally this binary may have been good enough, but not in light of the development of bioprediction. If we wish to avoid ending up like Nasrudin and King Lear, we must re-evaluate our allegiance to the fading utility of this division.

The concept of risk, I will argue, is the central component of disorder that has been obscured by historical efforts to make and solidify hard distinctions between disorder and normalcy. The challenges that biomarkers pose to this binary model illustrate the central role that likelihood and type of future harm play in theories of disorder, and that the dimensional magnitude of risk, not its presence or absence, is what matters. To drop the risk clause from the *DSM* would be a conceptual regression that could further widen the gulf between psychiatry and the rest of medical practice.

I propose that we should do the very opposite of this proposed change to the *DSM-5*; we should bring risk and probability from the shadowed corners of the concept of disorder to its center. I will call this reoriented theory of disorder the Probability Dysfunction theory of disorder. The remainder of this chapter is devoted to the development and defense of that theory.

Reorienting on an epicenter of probability allows a significant simplification of the existing theories of disorder, better classifies borderline conditions, makes explicit why other borderline conditions remain borderline, and permits the concept of disorder to be strengthened, not confused, by the development of bioprediction.

ORIENTING WITHIN THREE INFLUENTIAL CONCEPTS OF DISORDER

Before we reorient, we must first orient ourselves within the recent philosophy of medicine and the three concepts of disorder that have been especially influential. These are the concepts developed by (1) Boorse, (2) Wakefield, and (3) Gert and colleagues. *DSM-IV-TR* (the edition prior to *DSM-5*) has also been influential, but for our purposes can be thought of as a combination of elements from 1, 2, and 3. I introduce the three theories from the philosophy of medicine here, briefly, so that we can see how a probability-focused concept emerges and differs from them.

All of these concepts refer to the overarching concept of disorder; Boorse calls it "disease," Gert and colleagues prefer the less linguistically entrenched term "malady," and Wakefield explains: "some writers distinguish between "disorder, 'disease', and 'illness'; I focus on 'disorder' as the broader term that covers both traumatic injuries and diseases/illnesses, thus being closer to the overall concept of medical pathology" (Wakefield 2007, p. 150).

Like Wakefield, I will refer to the broad concept as "disorder." The distinctions between these different terms are important (Boorse 1975; Gert et al. 2006), but are also confusing; I will argue, however, that they can be derived from my overarching Probability Dysfunction concept of disorder as different types of "risk bands," both of which I will explain in this section. I will also operate on the assumption that this concept should apply equally in all realms of medicine, from infectious disease to oncology to psychiatry; to quote Wakefield again, "the claim of psychiatry to be a medical discipline depends on there being genuine mental disorders in the same sense of "disorder" that is used in physical medicine" (Wakefield 2007, p. 150).

The three theories are introduced below and are listed in Table 4.1. As I describe the theories in the following text, note that (1) Boorse and Wakefield do not have the risk clause, but Gert and colleagues do; (2) all deal with deviation from a norm, but Boorse and Wakefield hinge on deviation from natural function as designed by natural selection, while Gert and colleagues are teleologically agnostic; (3) Wakefield and Gert and colleagues incorporate a harm criterion while Boorse tries to stay "value free"; and (4) Wakefield's harm criterion is context relative while Gert and colleagues' is universal (see Table 4.1).

Boorse's Natural Dysfunction

> A disorder is a "deviation from the natural functional organization of the species" where the natural functions of a component are "its species-typical contributions to the apical goals of survival and reproduction" and "a [disorder] is an *illness* only if it is serious enough to be incapacitating, and therefore is (i) undesirable for its bearer; (ii) a title to special treatment; and (iii) a valid excuse for normally criticizable behavior." (Boorse 1975, p. 61)

Table 4.1 The Three Influential Theories of Disorder

Theory	Deviation from	Harm Criterion	Type of Harm	Risk Clause
Boorse	Natural function	No	n/a	No
Wakefield	Natural function	Yes	Relative	No
Gert and colleagues	Average function	Yes	Universal	Yes

Wakefield's Harmful Dysfunction

> A disorder is a "harmful dysfunction." Harmful is a value normative term and dysfunction is a failure of a mechanism to fulfill its natural function as designed by natural selection. (Wakefield 1992, 2007)

Gert and Colleagues' Harmful Abnormal Condition

> A disorder is "an abnormal condition, other than her (his) rational beliefs or desires, such that she (he) is suffering, or is at significantly increased risk of suffering, a nontrivial harm or evil (death, pain, disability, loss of freedom, or loss of pleasure) in the absence of a distinct sustaining cause." (Gert et al. 2006, 142).

The core of the Boorsian concept of disease is a (1) deviation (2) from the species-typical, (3) natural design (as produced by natural selection); if that deviation is judged to have certain adverse effects, we can properly call it an illness. Boorse claims that the deviation from the species-typical natural design is a value-free objective measure, and normative value judgment enters only at the level of illness.

Wakefield's concept is similar to Boorse's, in that it also operates on natural teleological dysfunction, but he qualifies that the dysfunction must be harmful to count as a disorder. For our purposes, it is sufficient to point out that the key difference between Boorse and Wakefield is that for Wakefield, the overarching term (disorder) is value-hybrid (containing both objective judgment of dysfunction and value judgment of harm).

The third concept, advanced by Gert and colleagues (2006), is also a value-hybrid concept, but differs from Wakefield primarily in the conception of harm. While Wakefield admits value-relativism in what counts as harm, Gert and colleagues restrict their concept to universal harms. Another difference, which will become important, is that Gert and colleagues state that the condition is abnormal, but remain agnostic on natural design.

Because I am particularly concerned with biomarkers in psychiatry, I also include the definition from the last version *DSM*; as an exercise, note that it has significant similarity to all three of the concepts introduced above.

The *DSM-IV-TR* Concept of Disorder

> [. . . A mental disorder is] a clinically significant behavioral or psychological syndrome or pattern that occurs in an individual and that is associated with present distress (e.g. a painful symptom) or disability (i.e. impairment in one or more important areas of functioning) or with a significantly increased risk of suffering death, pain, disability, or an important loss of freedom. In addition, this syndrome or pattern must not be merely an expectable and culturally sanctioned response to a particular event, for example, the death of a loved one. Whatever its original cause, it must currently be considered a manifestation of a behavioral, psychological, or biological dysfunction in the individual. Neither deviant behavior (e.g. political, religious, or sexual) nor conflicts that are primarily between the individual and society are mental disorders unless the deviance or conflict is a symptom of a dysfunction in the individual, as described above. (APA 2000)

Of particular relevance, as mentioned earlier, is that this *DSM-IV-TR* definition (but not the one in the *DSM-5* or the originally proposed revision to *DSM-5*) incorporates the risk clause. The *DSM-IV-TR* conceptualizes harm in a way very similar to Gert and colleagues' definition, but also incorporates the concept of dysfunction. It also has a range of exclusion clauses, the most important being "expectable responses" and non-dysfunctional conflict between individual and society.

In the proposed revision of the definition for the *DSM-5*, we can see a larger shift away from Gert and colleagues toward Wakefield:

> A mental disorder is a health condition characterized by significant dysfunction in an individual's cognitions, emotions, or behaviors that reflects a disturbance in the psychological, biological, or developmental processes underlying mental functioning. Some disorders may not be diagnosable until they have caused clinically significant distress or impairment of performance [. . .] A mental disorder is not merely an expectable or culturally sanctioned response to a specific event such as the death of a loved one. Neither culturally deviant behavior (e.g., political, religious, or sexual) nor a conflict that is primarily between the individual and society is a mental disorder unless the deviance or conflict results from a dysfunction in the

> individual, as described above. (*DSM-5* Task Force, APA 2012, http://www.dsm5.org/proposedrevision/Pages/proposedrevision.aspx?rid=465)

The risk clause is removed, as well as the explicit dependence on harm; it retains the exclusion of expectable responses and conflicts between individual and society. The dysfunction is dependent on a disturbance, though both are poorly defined.

The definition of disorder that ended up being published in the *DSM-5*, though slightly changed from this last proposal from the working group, shares the core philosophy championed by the working group through its removal of an explicit risk clause:

> A mental disorder is a syndrome characterized by clinically significant disturbance in an individual's cognition, emotion regulation, or behavior that reflects a dysfunction in the psychological, biological, or developmental process underlying mental functioning. Mental disorders are usually associated with significant distress or disability in social, occupational or other important activities. An expectable or culturally approved response to a common stressor or loss, such as the death of a loved one, is not a mental disorder. socially deviant behavior (e.g. political, religious, or sexual) and conflicts that are primarily between the individual and society are not mental disorders unless the deviance or conflict results from a dysfunction in the individual, as described above. (*DSM-5*, APA 2013, p. 20).

In the reorientation that follows, I will argue that

1. the concept of disorder cannot be dependent on natural function (a point others have argued [Gert & Culver 2004]);
2. it should be dependent on biological variation linked to harm in some relevant way;
3. the relevant way must incorporate a risk clause;
4. the exclusion clauses of (a) external sustaining cause in Gert and colleagues and (b) non-dysfunctional conflict between individual and society and of expectable responses in the *DSM* are unsupportable and unnecessary;
5. these exclusion clauses actually reflect concerns of justice and of the shape of the condition's harm-probability curve, both of which should be made explicit; and
6. showing the centrality of risk in disorder enables a synergy rather than a conflict with biomarkers.

INTRODUCTION TO THE "PROBABILITY DYSFUNCTION"

I present the skeleton of my probability-reoriented concept here as an example that incorporates the six points noted in the previous section. As I demonstrate where this model conforms to and departs from the other major theories, and provide justification for those departures, we will slowly add muscle to the skeleton; it is my hope that we will have added enough meat to it by the end that we can put the concept to work.

The Probability Dysfunction

> A disorder is a variation from a biological average (an "image average"), is associated with an increased probability of harm (present or future) after subtracting the effects of social injustice, and has a probability function that justifies a certain response or attitude.
>
> Subtypes of disorders flow from the stratification of magnitude of probability and type of harm.

Like the concepts described earlier, this one has both objective and value-judgment components. The objective components are (a) the variation, and (b) the association of that variation with a particular event. There are three normative judgments: (1) that the event is a harm, (2) that the increased probability is not due to injustice,[2] and (3) that the general probability function (incorporating individual increases and decreases in risk) associated with the variation is somehow bad (a "dysfunction") and justifies a response. The response may be treatment or the adoption of a certain attitude.

Now, let us put the skeleton together, with the purpose of clarifying conceptual disagreement about the ethical implications of bioprediction.

HARM, NOT NATURAL DYSFUNCTION, SHOULD PROVIDE DIRECTIONALITY TO VARIATION

All of the concepts of disorder incorporate variation between individuals, but the question is, variation from what?

Boorse, in his 1975 *Philosophy and Public Affairs* paper, identified that basing disorder upon variation from pure statistical means encounters difficulty: "many deviations from the average—e.g. unusual strength of vital capacity or eye color—are not unhealthy. [Then again,] practically everyone has some [disorder], and there are also particular [disorders] such as tooth decay and minor lung irritation that are nearly universal" (Boorse 1975).

This concern is partially about a *directionality* requirement of disorder: variations that in some sense function better or merely differently are not disorders. Boorse's solution (and Wakefield's) is that the variation must be variation away from function according to natural design. Since neither eye color, nor unusual strength, nor vital capacity impairs the species-typical contribution of the eyes, muscles, or lungs to the natural selection goals of survival and reproduction, they are not disorders.

Though natural teleology seems at first glance to imbue the right type of directionality, it ultimately fails because of uncertainty in natural function[3] and the possibility of programmed deterioration. Gert and colleagues succinctly sum up: "the claim that dysfunction is a failure of nature's design is often unverifiable. Perhaps nature designed people to deteriorate and die to allow for the species to develop" (Gert & Culver 2004). I will only briefly illustrate these arguments because they appear elsewhere (Cooper 2002; Gert & Culver 2004) and are secondary to our concerns about bioprediction.

Boorse has the intuition that tooth decay and lung irritation are disorders, not phenomena that are in the "nature of the species." But we cannot conclude that teeth, lungs, and other components of biology are not designed by natural selection to perform a function only within a certain time frame, for example, long enough to have and rear children. Many biological processes "wear," as a rechargeable computer battery does.[4] Though with every recharge of my computer battery, the battery stores less charge, we would not say that it is no longer performing its designed function, especially if I expect to transfer all my data to a new computer in four years.[5]

Because many different evolutionary strategies could be "good enough," the species could tolerate an equally diverse range and quality of "battery life." Natural dysfunction, as one might expect, becomes particularly confusing when we consider *geriatric disorders*, including dementia. The enterprise of trying to argue that one condition is either a natural function or a dysfunction leads to enormous controversy and confusion, and what is gained in this debate is often unclear. What if we did discover that all geriatric disorders are examples of natural functioning as designed by natural selection?

That we would be more likely to adjust our concept of disorder than to stop referring to these afflictions of the elderly as disorders suggests that something else is driving the appeal of teleology. A "failure of a mechanism in the person to perform a natural function for which the mechanism was designed by natural selection" (Wakefield 1992) produces a strong intuition that this change in function is likely to be harmful either now or in the future. It is the constancy of the association of geriatric disorders with harm that supports their inclusion in a concept of disorder despite controversy about programmed deterioration. Conversely, the absence of an association with present or future harm drives the exclusion of a range of dysfunctions from the domain of disorder, one example being the temporary sterility due to taking oral contraception (Cooper 2002, 265).[6]

Wakefield's concept of a *harmful* dysfunction excludes Cooper's non-harmful dysfunctions. His inclusion of teleological dysfunction still gets the concept into trouble when trying to evaluate geriatric disorders, however, which are harmful but not clearly teleological dysfunction.

It is the harm criterion,[7] then, that should satisfy the unidirectional requirement of disorder. Gert and colleagues, like Wakefield, appeal directly to harm as important in the concept of disorder, but they wisely distance themselves from teleology and thereby avoid its drawbacks. Such a strategy excludes conditions like abnormal strength and contraceptive-induced sterility, but allows the inclusion of geriatric disorders.

VARIATION FROM AN IMAGE AVERAGE AND INCREASED RISK OF HARM

We must say something more about the kind of functional biological variation we have in mind, if we are to remain agnostic about natural function (and thus variation from it). In my definition, I use variation from an *image average*.

Image averaging is the type of averaging that goes on when a structural biologist tries to resolve a crystal structure; many different crystals of the same protein, each with slight individual variations in conformation, are aligned and then averaged together. This might be explained analogically if we were to take 1000 heart rhythms then align and average them together, giving weight to each feature in the rhythm according to its frequency within the 1000 rhythms. Infrequent variations would drop out of the image average and you might end up with what would approximate the heart rhythm found in a textbook. This image average heart rhythm would be an extreme statistical outlier in the population, if it is indeed possessed by someone at some point in time, and thus the image average diverges from Boorse's conception of the species-typical (see Murphy 2008). This type of image average is also similar to what is used to establish the reference or template brain in fMRI and CT brain imaging studies, and to the idea of a reference genome in genomic studies.

The first strength of the image average is that it provides a teleologically agnostic reference for variation. The second strength is that it functions in a world where most people have some type of variation from the image average—even *many* variations—as long as the variations are different, which becomes important when we return to Boorse's common diseases. As long as tooth decay is not present in the same location in most members of the species, it will drop out of the image average and count as variation (and thus a candidate for disorder).[8]

The third strength of the image average is a practical one, in that it might be approximated in medical research with a *normal experimental control*. The experimental control is an average on some measure of a number of representative specimens.[9] We can integrate the philosophical concept with epidemiological studies (like the Framingham Heart Study) that examine whether a particular variation (or set of variations) is associated with harm. We can see, also, that this reference allows a logical interface with biomarkers.

In the Probability Dysfunction theory of disorder, therefore, linking variation from the image average with the concept of harm provides directionality. If unusual strength, eye color, and vital capacity are not associated with harm or an increased risk of harm, then they are not disorders. If tooth decay and lung irritation are associated with harm or increase the likelihood of future harm, then they are candidates for disorders. If contraceptive-induced sterility is not associated with harm or increased risk of harm, then it is not a disorder; conversely, if that particular contraceptive-induced sterility leads to a significantly increased risk of harm, for example high risk of pain and death due to ovarian cancer, then it may be a candidate for disorder. Moreover, if there is an increased risk of harm without variance from the image average, as is the case when a person drives a car, then this is not a disorder.

A final strength of the biological image average is that it is relatively robust to a general objection to averaging: how individualized to make the image averages. The flavor of the objection is that if we are too general in our averaging, everyone is highly variant; if we are too individualized, no one is. To illustrate, consider the alarmed medical student upon hearing for the first time the incredibly high heart rate and the whoosh of a murmur in the heart of an infant. These are variants that in the adult are associated with increased risk of harm, but the medical student will learn that in the infant these features are neither variant nor associated with risk of harm. The biological image average, I think, need only take into account broad-strokes biological development; the fine tuning of what counts as a candidate for disorder might be achievable by this rough picture of development in combination with the harm criterion, just as the harm criterion provides directionality.[10] One such developmental average with which many will be familiar is the growth curve: certain variations are associated with increased harm (e.g., in the *shape* of or rate of change in the curve), while others do not have this same increase (e.g., a left- or right-shift of the *position* of the curve). Biomarkers might be informative here insofar as they help us estimate the (a) likelihood of deviation from the rough developmental average, and (b) the likelihood of that deviation being associated with harm or increased risk of harm.

THE IMPORTANCE OF THE RISK CLAUSE

Why is it necessary to talk in terms of increased risk? After all, as Gert and colleagues point out, "most others fail to consider [the risk clause]" (Gert et al. 2006, 135). In explaining why their concept of disorder incorporates the risk clause, Gert and Culver offer the following examples:

> High blood pressure is a physical disorder that may involve no present distress or disability but does involve a significantly increased risk of death [. . .] Anorexia nervosa is associated with a significantly increased risk of death. (Gert & Culver 2004)

I will go further to argue that the risk clause enables one to deal with several types of conditions that would otherwise be problematically excluded: (1) chronic but managed conditions, (2) periodic conditions, and (3) conditions with signs or symptoms that are not themselves especially distressing (and would be ignored by some), but that increase risk of serious adverse events.[11]

Chronic, Managed Conditions

Cooper (2002, as discussed in Murphy 2008) has argued that people with chronic conditions that are managed well by drugs function normally, but nonetheless still have disorders. Cooper argues that natural functioning accounts, like Boorse's, must at the least talk about increased risk of dysfunction. Without the risk clause, moreover, well-managed allergies, phobias, herpes, and so on, would not qualify as disorders by Boorsian/Wakefield accounts until they ceased being controlled. Theories that are teleologically agnostic also require a risk clause to include chronic but managed conditions.

Periodic Conditions

The harmful symptoms of many disorders are periodic. Examples are relapsing-remitting multiple sclerosis, herpes, epilepsy, bipolar disorder, angina, the memory lapses in dementia, the attacks in sickle cell anemia, and hypo- and hyperglycemic states. The increased likelihood of infections in acquired immune deficiency syndrome (AIDS) is another example. If there is no present harm between cycles, there would be no disorder according to concepts without a risk clause (i.e., Wakefield, Boorse, *DSM-5*). A concept that incorporates the risk clause is necessary to include these disorders.

Conditions with Symptoms That Are Not Especially Distressing, But Increase Risk of Adverse Events

Many conditions may have noticeable signs or symptoms that are not themselves particularly distressing at the present time, but those signs/symptoms are worrisome primarily because they signal an increased risk of serious harm in the future (see Table 4.2 for examples). Hemophilia, for example, puts people

Table 4.2 Examples of Disorders with Increased Probability of Future Harm (But Mild or Unnoticeable Present Harm) Already in Use*

Disorder	Increased Risk of
Osteopenia	Bone fracture
Osteoporosis	Bone fracture
Hypertrophic cardiomyopathy	Sudden cardiac death
Hemophilia	Profuse bleeding leading to pain or death
High blood pressure	Stroke, heart attack, embolism
Hyperlipidemia	Stroke, heart attack, embolism
Dormant herpes	Pain, fever, ulcer
Immune deficiency	Infection-related harm
Adenomatous polyposa	Stricture of gastrointestinal tract, colon cancer
Marfan syndrome	Aortic rupture-related sudden death, blindness (detached retina)
Diabetes	Coma, death (hypoglycemia), ketoacidosis, blindness (retinopathy), kidney failure (nephropathy)
Anorexia	Death, infertility
Werner syndrome	Bone tumors
Familial Mediterranean fever	Fever
Familial cold urticaria	Inflammation
Bloom's syndrome	Cancer
Rhabdoid predisposition syndrome	Rhabdoid tumors
Atherosclerosis	Stroke, heart attack, pain and loss of function in kidneys, lower extremities
Cancer	Pain, death
Human papilloma virus (HPV)	Cervical cancer
Renal insufficiency	High blood pressure; kidney failure

*details on most of these disorders can be found in Kumar et al. (2010)

at high risk for uncontrolled bleeding and death. Osteoporosis puts people at increased risk of fracture. Hypertrophic cardiomyopathy (thickening of the heart muscle) puts people at risk of sudden cardiac death (often young, athletic people). Infection with human papillomavirus (HPV) often goes unnoticed but increases the risk of cervical cancer, which in turn increases risk of death[12] (Kumar et al. 2010). Without buying into the proposition that deviation is intrinsically bad, it seems as if Wakefield would be committed to argue that a person with hemophilia would have a disorder only during the act of bleeding to death.

A theory of disorder needs to incorporate a risk clause, in order to include chronic but managed conditions, periodic conditions, and conditions with minor signs/symptoms but significant probability of future harm. I shall show in the next section that estimation of future harm plays an even more fundamental role in disorder than this discussion suggests.

PREDICTION OF *FUTURE HARM* IS THE CONCEPTUAL CORE OF DISORDER

By remaining teleologically agnostic, providing directionality through linking harm to variation, and incorporating the risk clause, Gert and colleagues' definition of disorder avoids the pitfalls I have so far named; besides my more specific characterization of variation from the image average, rather than Gert and colleagues' general "abnormality," their definition to this point closely resembles my Probability Dysfunction. My definition differs in two important ways: (1) the degree of emphasis on probability of future harm, and (2) the substitution of the concept of justice for Gert and colleagues' "distinct sustaining cause."

This section aims to show that the *probability of future harm*, as estimated by past and present signs, symptoms, and other harms, *is actually more central to the concept of disorder* than are present harms. Bioprediction, insofar as it is used to estimate the probability of future harm, consequently should work synergistically with a concept of disorder by performing a role similar to classical clinical signs.

Consider Mrs. P:

> Mrs. P comes into your medical clinic for her 5-year physical exam. She reports being generally in good health, and that her only current medication is a daily multivitamin. When Mrs. P returns with a urine sample, however, she reports that she felt a burning sensation upon urination for the first time. A bit puzzled but concerned, you send the sample to the lab for urinalysis.

What are you hoping to learn from the urinalysis? The obvious answer is that you are hoping to learn whether Mrs. P has an undiagnosed disorder. But if we remember our discussion with Nasrudin and King Lear, we know that there is little value in a name alone. You might therefore agree with the *DSM-5* working group that a diagnosis of disorder "should help clinicians to determine prognosis, treatment plans, and potential treatment outcomes for their patients." You are hoping to learn whether the burning sensation is likely to return or whether it signals an increased likelihood of future harm like loss of a kidney, and how the likelihood of future harms might be altered by a certain course of action.

The concept of disorder is essentially forward-looking; the less support for increased risk of future harm, the less likely this burning represents the presence of a disorder now. If there is good evidence that the burning sensation is a one-time event, Mrs. P might have experienced a self-resolving disorder but is no longer experiencing one.

The presence of signs of disorder or urine biomarkers, however, is only one source of this forward-looking information. If the urinalysis is non-informative, but Mrs. P returns one week later with the report of three subsequent burning sensations, then the repeat might itself act as a marker of increased probability of future harm. We can use Bayesian-style reasoning to estimate that prior occurrence increases the likelihood of future occurrence. We make the assumption that past harm is predictive of future harm.

The existence of present harms, too, can play this predictive role. Disorder is dependent on the affirmative prediction that whatever harm is currently happening at t_1 has a high probability of recurring at some future time t_2, which may be separated from t_1 by a very small temporal increment, x (i.e., $t_2 = t_1 + x$). As this increment approaches zero, $\lim_{x \to 0}(t_1 + x)$, the recurrence of the harm through time will be experienced as the continuation of a present harm.

We might usefully think of disorder, therefore, in terms of probability of harm (y axis, dependent variable) as a function of time (x axis, independent variable), the plotting of which would create something very similar to a Kaplan-Meier curve.[13] To develop this graph, we need a set of markers of variation from the image average that is associated with an increased risk of harm; these markers can be traditional clinical signs or symptoms, but could also be predictive biomarkers (or a combination thereof).

To translate to *DSM* terminology and further tie in bioprediction, the past or present "distress (e.g. a painful symptom) or disability (i.e. impairment in one or more important areas of functioning)" (APA 2000) imperfectly predicts future occurrence. In psychiatry, where urinalysis has not been able to provide useful forward-looking information, disorders have existed on this symptomatic level. The recurrence or persistence of a set of symptoms for a defined period of time contributed to the prediction of future harm. Predictive biomarkers should be treated as different methods of making the same prediction.

Once we make the shift to think of disorder graphically as a function of time, we can see that hidden judgments about the shape of that curve underlie some of the controversy surrounding what counts as a disorder.

FLAWED EFFORTS TO EXCLUDE QUICKLY RESOLVING CONDITIONS: DISTINCT SUSTAINING CAUSES AND EXPECTABLE RESPONSES

Both Gert and colleagues and the *DSM* champion exclusion clauses that I argue are untenable in light of research on intersections of biology with environment; however, both clauses implicitly capture an intuition that quickly resolving conditions should be excluded from disorder, an intuition that is best made an explicit target of debate.

Gert and colleagues argue that a disorder can only exist in the *absence of a distinct sustaining cause* of the harm or of the increased risk of harm. According to the authors, a distinct sustaining cause is a cause that is separate from the person and "not only brings about an effect, [. . .] but whose removal results in the removal of the effect" (Gert & Culver 2004, 419). An embedded piece of shrapnel or integrated herpes virus, being difficult to remove, would be non-distinct sustaining causes.

Reliance on the presence or absence of a distinct sustaining cause becomes obviously problematic when we know about the interactions of biology and environment. Take the case of allergies or of PKU (reviewed in Chapter 2), for example, which intimately tie a biological variation with environment. Gert and colleagues argue (2006, 147) that an allergy restricts the person's freedom (e.g., to eat at certain restaurants), and that the presence of an allergen is not actually a distinct sustaining cause for this restriction (even though it might be the distinct sustaining cause for the harm when the person is exposed to the allergen). The case becomes even more counterintuitive when we consider increases in risk of harm (or loss of freedom), as it is not at all clear why the removal of the allergen from all environments would not eliminate this increase in risk (and thus count as a distinct sustaining cause).

Why do Gert and colleagues adopt this problematic formulation? It appears they do so primarily to exclude quickly resolving disorders, not phenomena with external sustaining causes per se.[14] Gert and colleagues explain, for example, that coughing upon leaving a smoke-filled room (i.e., upon the removal of the external cause) may count as a disorder if "the coughing continues for a significant amount of time, such as several hours, depending on its severity

[. . .].” However, “someone who coughs for only one or two minutes after leaving a smoke filled room is said not to have a [disorder]. There is little point in so labeling her momentary condition” (Gert et al. 2006, 150–151). Even though the distinct cause has been removed and the cough in the first case and the second are identical for those first few minutes, what seems important to Gert and colleagues is that the condition is likely to exist for some amount of time into the future.

This desire is similar to my emphasis on estimates of future harm in the concept of disorder, with the importance of present harm diminishing as the present approaches the instantaneous. In the brief-cough case, the probability curve has a steeply declining slope. Elsewhere, Gert and colleagues talk of the pain and disability (and risk of them continuing in the future) induced by being in a wrestler’s hammerlock as another instance of a distinct sustaining cause (Gert et al. 2006); though there would presumably be markers of present pain (e.g., release of prostaglandins), the biomarkers of increased risk of future pain would be absent. Focusing on markers of future harm rather than harm at the moment would do the work of the absence of a distinct sustaining cause in a way that permits environmental interactions.

But are we justified in excluding quickly resolving conditions? The cough, after all, could usefully be thought of as a self-limiting or resolving disorder. By making the shape of the probability function explicit, we have the option of stratifying types of disorder based on that shape.

While Gert and colleagues seem to wish to exclude disorders that self-resolve over minutes, the *DSM* definition effectively excludes some conditions that last weeks to months. The condition “must not be merely an expectable and culturally sanctioned response to a particular event, for example, the death of a loved one.”

The distress, pain, disability, and so on, however, from having a tree fall on one’s arm is both expectable and culturally sanctioned, but the broken arm is still a disorder. It may be a *self-limiting* disorder, but it is still a disorder. Similarly, the harmful symptoms (and increased probability of future harm) associated with post-traumatic stress disorder (PTSD) are an expectable response to the events endured by soldiers in war (and arguably, increasingly culturally sanctioned), but *that it is expectable* seems weak grounds for its exclusion. Moreover, as predictive biomarkers improve, and the domain of the “expectable” is expanded, maintaining such an exclusion seems untenable.

This particular exclusion in the *DSM* has recently come under fire in a debate about whether the depression-like symptoms following bereavement should count as a disorder (Stein et al. 2010). While I will not explicitly enter that specific debate here, I note that there is more agreement that depression-like symptoms that persist for a *very long time after* bereavement may qualify as a disorder. The strongest reasoning for such a division, like the case of the smoky room, seems to be a desire to impose a cutoff on self-limiting conditions.

In their influential 2010 editorial, Stein and colleagues also draw a dichotomy between self-limiting conditions and mental/psychiatric disorder:

> Although it may be useful for clinicians to distinguish between common responses to stressors and losses (*that are distressing but likely to be self-limiting, without high risk of persistent clinically significant distress or impairment*) and mental/psychiatric disorders (as defined here), common distressing reactions to common stressors and losses do carry an incremental risk of complications, including the development of mental/psychiatric disorders. (Stein et al. 2010, my emphasis)

Constructing such a strict dichotomy between self-limiting conditions that are common responses to stressors and disorder excludes the domain of twisted ankles, minor burns, and acute inflammation, which are common responses to stressors, but usually resolve without future risk. It would also exclude an open wound, but point out that it carries an incremental risk of becoming infected and converting to a different harm probability curve. Failure to resolve this tension may contribute to the strange recommendation, made later in the same paper, to remove the risk clause from the *DSM*.

Once we embrace the probability-grounded concept of disorder, and plot these controversial disorders as probability of harm as a function of time, we should see that what they all have in common is a *decreasing probability of harm over time* that is *expected* to resolve within a certain time window. One strength of the Probability Dysfunction concept of disorder is that it makes this time/risk relationship (and any cutoffs imposed by theorists) explicit.

Rather than exclude quickly resolving disorders, we should make the shape of their association with risk of harm over time explicit, and thereby differentiate them from more temporally extended disorders. Making this time relationship explicit redirects the conversation to a more fruitful path by forcing us to consider whether the apparent large difference in the length of this "cutoff" between psychiatry and the rest of medicine is justifiable, or merely reflects a historical absence of a certain *kind of disorder*, namely, a harmful but "normal" or self-limiting one.

RETURNING TO THE RISK CLAUSE, AND ITS DISAPPEARANCE FROM THE *DSM*

One lens into the tension created in psychiatry by the development of better methods to estimate risk of future adverse events is a case study of the revision of the *DSM* leading up to the *DSM-5*. The *DSM-5* Task Force dropped the risk clause from the definition of mental disorder in the last draft revision: "The phrase on risk is omitted in order to differentiate more clearly between disorders and risk factors"[15] (*DSM-5* Task Force, APA, 2012). The committee explicitly references Stein et al. (2010) as the basis of this decision.

In Stein et al. (2010), the authors explain that they recommend dropping the risk clause from the *DSM* in order to emphasize the difference between risk factors and disorders. I reproduce the entirety of the argument here. For reference, the precise phrase upon which they comment is also reproduced:

> *DSM-IV-TR*: [. . . associated with] *a significantly increased risk of suffering death, pain, disability, or an important loss of freedom*. (APA 2000)
>
> Regarding the phrase pertaining to "increased risk," risk factors are important to bear in mind and perhaps even to treat (indeed, the full title of the ICD-10 is International Statistical Classification of Disease and Related Health Problems, with the latter phrase including risk factors for disease such as hypertension); perhaps DSM-5 should consider an analogous extension to its title. A full consideration of this issue is beyond the scope of this editorial; diagnosis and treatment of risk factors for psychiatric disorders is appropriately a contentious area, where advantages and disadvantages must be weighed carefully. At the same time, we would note that disorder and risk factors should not be conflated [. . .] We therefore tentatively suggest simplifying this criterion by omitting the phrase on risk [. . .]. We recognize, however, that limiting the classification to deal with disorders only may be unduly restrictive. (Stein et al. 2010, p. 1762)

While it is difficult to discern Stein et al.'s arguments for the difference between risk factors and disorders, one educated estimation is that it lies in the differential magnitude of association with harm assumed in the terms "risk factor" and "disorder." While it may previously have been good enough to assume a dichotomy of "low-probability of harm risk factors" and "high-probability disorders," the further development of predictive biomarkers and their multiplexing into actualial risk predictions makes maintaining such a dichotomy untenable (as we could have high probabilities from collections of low-probability risk factors). If the risk clause is even a fraction as central to the concept of disorder as I have argued here, dropping it from the *DSM* without stronger argument is premature.[16]

The redevelopment of disorder to a risk/time concept provides a framework by which we can usefully categorize differences in the probability of harm over time. Historical distinctions between "common responses," or what Stein and colleagues call risk factors and disorders, should be reconceptualized as markers with very different magnitude risks of harm, and should be separated on that basis.

Practically, how might we develop this stratification? Though not a complete solution, one useful metric would be the area under the curve (AUC) in our probability over time graphs (i.e., the integral of the probability function over a particular interval); this would help separate self-limiting conditions as well as give us a measure to compare high probabilities over short time windows, and lower probabilities over longer time windows.

THE PROBABILITY FUNCTION AS LANDSCAPE

Until now, we have operated with "harm" as a very general term. There are many different types of harm, of course, such as pain, death, disability, loss of freedom, loss of pleasure, etc. Each of these "general harms" is composed of a set of specific harms, such as loss of control of voluntary movement, which itself could be broken down into bradykinesia (slowness of intended movement) and dyskinesia (appearance of involuntary movement) and so on. One possible way of using the Probability Dysfunction concept to capture this complexity is to construct probability functions (our probability over time graphs) for each general harm, and for common specific harms. Doing so would create a large spreadsheet of risks very quickly. This is not necessarily a problem, as tools for handling large data sets (i.e., Big Data) have immensely improved. A lot of computation could be done automatically and "risk clusters" identified for each patient. This risk clustering, because it cuts across existing disorders, is similar to problem-based clustering (Filiou & Turck 2011) or symptom clustering (reviewed in Hyman 2011).

We identified in Chapter 3 that one of the challenges of bioprediction is that it makes explicit that variation can change the probabilities of many individual harms simultaneously, a concept introduced as pleiotropy. One aspect of pleiotropy that could be captured in the Probability Dysfunction concept that is hard to model in traditional systems is when variations from the biological image average might be associated with an increased risk of one type of harm and a *decreased* risk of a different harm; the further specified the harms, the more likely that this would be the case. For example, an underactive immune system might have an increased risk of opportunistic infection and the resulting harms, but a *decreased likelihood* of the harms associated with autoimmune disease. Low clotting factor in blood could increase the risk of harm due to hemorrhage but decrease the likelihood of harm associated with internal blood clots. Variations that increase the likelihood of multiple specific harms and decrease the likelihood of others might be lost in traditional disorder categories but could be taken into account statistically in a probability dysfunction alongside integrated risks for general harms.

We might think of a *particular* disorder as the sum of a series of three-dimensional wave functions of risk of specific harms over time. With the right computational tools, both the summation and the individual component waves could be modeled.[17] Clinical risk calculators, such as the Reynolds Risk score for cardiovascular events, might be viewed as early examples of what individual dimensions could look like (see Table 4.3). These risk calculators sum a large range of sources of risk information to estimate the probability of a specific harm over a particular time window.

Now we have constructed a more holistic representation of probability of harms, where there are peaks (increased risks) and valleys (decreased risks).

Table 4.3 Sample Risk Calculation Scores

Risk Tool	**Description**	**Website**
Reynolds Risk Score	10-year risk of heart attack, stroke, or other major heart disease	www.reynoldsriskscore.org/default.aspx
BRCAPro	Risk of having BRCA1/2 breast/ovarian cancer genes	bcb.dfci.harvard.edu/bayesmendel/brcapro.php
Framingham Risk Scores	• Atrial Fibrillation (10-year risk), • Cardiovascular Disease (30-year risk), • Congestive Heart Failure, • Coronary Heart Disease (10-year, 2-year risk), • Diabetes Risk Score, • General Cardiovascular Disease, Hard Coronary Heart Disease (10-year risk), • Hypertension Risk Score, • Stroke, • Stroke after Atrial Fibrillation, • Stroke or Death after Atrial Fibrillation	www.framinghamheartstudy.org/risk/index.html
WHO Fracture Risk Assessment Tool	Bone fracture	www.shef.ac.uk/FRAX/tool.jsp?country=9

It is this overall landscape to which I refer, as a heuristic, as the *probability function*. This function should be thought of as dynamically changing as the environment changes, as different biomarkers appear and disappear, and as time flows. In order for this probability function to count as "dys"-function, we need to make a judgment that the degree of risk of harm, the type of harm, and time period of high risk are sufficiently "bad" in my likely environments as to warrant a certain response. The paradigm is a systematized generalization of how Karlawish (2011, pp.1488) describes cardiovascular disease: "cardiovascular disease is defined as the probability of a cardiovascular event that is sufficiently high that a clinician *ought* to recommend interventions to reduce that risk" (my emphasis, Karlawish, 2011, pp. 1487–1493).

Deciding the probability thresholds at which it is appropriate to make the judgment that a probability function is a probability dysfunction that someone "ought to" respond in a certain way is where I see much of the hard labor of practical neuroethics of biomarkers to lay, and I make preliminary steps to discuss it in Chapter 6. The response warranted for each risk stratification may be a type of medical treatment, moreover, but, especially at the intersections of disorder and other social institutions, may be a certain attitude in judgments of need or moral responsibility.

We might think of medicine as an elaborate exercise in risk management. We already think in terms of risk-benefit analysis in treatment, and that the degree of acceptable risk changes with the severity of disorder. The Probability Dysfunction concept enables us to see the disorder itself as a probability of harm. Treatment is justified only under the assumption that future occurrence of harm is probable, and the type of treatment is matched to that risk with the goal of reducing the probability of future (re)occurrence of harm.

To help us make decisions about treatment, we might usefully model treatment as "adding in" another probability wave into the probability landscape. The presence of a drug at a certain level in the body might itself be thought of as a biomarker of risk; it increases risk of some specific harms (the "side effects") and decreases the probability of other harms. The variations from the image average that increase risk of harm associated with vaccination (soreness, risk of encephalopathy) or surgery (continued pain, risk of infection) might be thought of as wave functions with short-term, self-resolving bounds that, when summed into the personal probability function, reduces other risks. Treatment, therefore, will often be risk *redistribution*.[18] It is at this level that the etiology of disorder becomes very important.

We might anticipate, however, that levels and types of risks might be interpreted and responded to differently in the legal and political context. In the legal setting, for example, we might be particularly concerned with the probabilities, not of harm per se, but of specific events that are relevant to moral responsibility, and risk might warrant a mitigation or enhancement of responsibility. The political sphere might be concerned with still different probabilities, for example, those relevant to social justice. I return to these ideas in Part II and Part III.

One objection likely to be raised to adopting the Probability Dysfunction concept is that it seems to have far too many moving parts to be practically approximated. Since this risk landscape might constantly change as my biology changes, as my environment changes, and as time flows, how can we graph such a dynamic function, let alone use it to inform actions?

Unless we are constantly measuring variations and updating the function in real time, we are likely to remain ignorant of this sort of rapid fluctuation, and thus the practical dilemma is how to usefully manage this temporal uncertainty. Though these inputs to the risk landscape change, we will often be able to estimate the *likelihood of change* to a fine enough degree to be practically relevant. We have actually already encountered this way of managing uncertainty in regard to biomarkers when we considered that some biomarkers might be associated with increased risk over a short time window and others might be associated with a risk of similar magnitude but distributed over a much longer time window. In essence, these two types of biomarkers encode temporal likelihoods. We might also have uncertainty in the degree of confidence we should place in the prediction associated with a set of biomarkers, for example, if there is larger individual to individual variation with some biomarkers than others; in this case, too, we could weight the magnitude of the risk prediction according to the likelihood that it is accurate. Similarly, we might know that a certain biological variation, for example a bee allergy, has a relatively constant risk relationship across time but is highly dependent on environment (e.g., highest at an apiary). This knowledge might be modeled in the particular probability function by weighting the risk of harm of that biological variation in each environment, according our best estimates of the likelihood of encountering that environment in a given time window (e.g., much lower during winter than spring and summer). As we assay biomarkers or have changes in the likelihoods of environments, we could update the probability function, just as we update the fracture risk score each time we do a bone density scan.

RISK BANDING

A second practical objection is that it may be too much to ask of society to shift directly from the current naming system to probability functions in a graphical interface. While one day we may become more comfortable thinking of disorder directly in this probabilistic format, a practical half-step would be to try to reconstruct existing nosology in a way that more accurately captures the magnitudes and shapes of risks. Reworking existing structures (in this case, nosology) in a way that groups similar risks together in a way that facilitates treating like risks alike is an application of a practical strategy that I call careful *risk banding*. Careful risk banding is accomplished more satisfactorily in the nosology of preventive cardiology than in that of many other fields: To return to the Karlawish example, the name "cardiovascular disease" provides heuristic information on the magnitude of risk of a specified harm over a given time window, which separates it from conditions of lesser cardiovascular risk and guides specific response, in this case particular treatments. Similarly, doctors are in a sense risk banding when they calculate a 5-year risk of hip fracture and refer to a range of slightly elevated risks as osteopenia and a range of higher risks as osteoporosis, and recommend different actions accordingly. The process of risk banding simplifies a probability dysfunction along a salient axis, and consequently one trade's completeness of information about that probability function for practical utility. Doing so *well* can save time and can speed communication; doing so poorly (as is the case with much existing nosology) can obscure rather than highlight the relevant risks.

Once one realizes that different names *already* act as heuristics about risk, it will be readily recognized that disagreements about the appropriateness of inclusion in a certain diagnostic category often *actually* reflect different hidden assumptions about the size and shape of risk that a particular label should (or does) represent, that is, whether those two conditions belong in the same *risk band*. In response to the World Health Organization's (WHO) recent report on the burden of depression, for example, Wakefield objects that the "extreme degree of severity [of burden the WHO attributed to depression] assumes that all cases of depression share the depth, chronicity, and recurrence that are characteristic of the severe conditions that health workers see in their practices" (Wakefield 2007, p. 153). It is the differing *magnitude and shape* of the risk of future occurrence of harm in the conditions included that seem to be at the core of Wakefield's complaint. If depression-like symptoms carry a different risk of future harm (e.g., recurrence) if they occur in association with the behavioral marker of "consulting a health worker" than if detected in a survey (which Wakefield posits might include those with depression symptoms after bereavement), then they should not be grouped together.[19]

Rather than dropping the risk clause from the concept of disorder, therefore, a fruitful approach could be to emphasize careful *risk banding*—that is, banding together with like terminology sets of variation with like magnitudes and types of risk of harm over time. Again, such a solution is very imperfect, as we risk losing sight of the probability landscape in which we wish to capture isolated features. However, grappling with how risk banding could be carried out successfully often enables one to better understand some otherwise opaque debates within health. I will address a practical example of careful risk banding in Chapter 5.

While debate on where to set the thresholds for different bands (and how many bands to have) will be a difficult practical problem, it is a debate that we should consider directly. Biologically based risk estimation will only improve with time, and we will be increasingly unable to ignore the information about risks of future harm that such estimation will provide.

JUSTICE SHOULD BE EXPLICITLY LINKED TO DISORDER

Without the clauses of absence of a distinct sustaining cause and teleological dysfunction, the Probability Dysfunction definition of disorder is susceptible to a different kind of challenge: What happens if some of the increased probability of harm is due to discrimination, say, based on variation in skin pigmentation?

One aim of including natural function or distinct sustaining causes is to exclude biology that triggers persecution (i.e., harms due to racism) from the domain of disorder. A teleological dysfunction account could say that the color of one's skin, for example, even if associated with increased risk of harm due to discrimination, is not a natural dysfunction and therefore not a disorder. Gert and colleagues could point to the discriminative environment as a distinct sustaining cause, and thereby disqualify this case from disorder. But since we have seen that there are deep problems with using natural functions and distinct sustaining causes, we need to look elsewhere in order to exclude the interaction of skin color and discrimination from the realm of disorder.

I propose to do so by openly acknowledging the intersection of the concepts of disorder and justice. Rather than appeal to natural functions and distinct sustaining causes, since we have found both these strategies to be deficient, a more honest, and hopefully productive, approach is to openly debate the degree to which a biological variation's association with increased risk of future harm is due to an *unjust society.* If the increased risk is primarily due to an injustice (for example because of wrongful discrimination based on the variation), then the condition is not a disorder.[20] If, on the other hand, the increased risk is not due to an injustice, then the condition would be a candidate for disorder; because we do not view as unjust an increased risk of harm due to living in societies that commonly eat peanuts or foods high in phenylalanine, then peanut allergies and phenylketonuria would still count as disorders.

Though it is controversial to explicitly include justice in a definition of disorder, I think doing so better enables a productive discussion of conditions that may have risks of harm partially due to injustice.

One condition that is on the edge, for example, might be a biological predisposition to impulsive violence (see Chapter 7). If this biological predisposition leads to an increased likelihood of being incarcerated, and if we do not view a society that incarcerates violent criminals *in general* as unjust, then this biological predisposition to impulsive violent crime might qualify as a disorder. Or Gert and colleagues raise the example of a condition, a physical abnormality, which increases the probability of harm due to society's reaction (Gert et al. 2006, 148). The question is whether that reaction is unjust, not whether it is expectable, normal, or natural.

NODES OF MORAL DEBATE MADE EXPLICIT

In this reconceptualization of disorder as a probability dysfunction, I have pointed out nodes of normative controversy. These nodes are (1) what counts as harm, (2) where to set risk bands and whether there is a threshold below which something ceases to be a disorder, (3) whether an increase in risk of harm is due to injustice, and (4) what response is justified. This reconceptualization is not meant to resolve this controversy, but to make these nodes explicit so that we can bring them within the sphere of clear and reasoned normative debate.

Gert and colleagues try to minimize the controversy at the harm node by restricting their analysis to what they call universal harms: pain, death, loss of freedom, loss of pleasure, or disability. But even here we can see that debate is likely to arise about the specifics of each of these; there is perhaps a more substantial literature on what counts as a disability, for example, than on disorder (Savulescu & Kahane 2011).

Rather than try to restrict the domain of harm in this work, I submit merely that in order to qualify as a disorder, a case must be argued that the event associated with the variation from the image average is in some way harmful. To illustrate how the Probability Dysfunction concept can help us think systematically about even the most ambiguous cases, consider the following case: If a person has a substantial unconscious racial bias, but strong egalitarian commitments, might her substantial unconscious racial bias qualify as a disorder? We can use the nodal framework in the Probability Dysfunction theory to systematically organize our contemplation of this very unclear question. We return to the nodes of controversy, here, harm and justice; the person has a variation from the image average that increases the likelihood that she will behave in a discriminatory manner. Is this harmful? One might put forward the claim that she may have distress or pain from her committing an act that runs contrary to her values, and she may be disciplined by her community if she indeed does discriminate. If, after further discussion, we decide that a society that punishes discrimination (in general) is not unjust, then the person's initial unconscious bias may rightly count as harm to her. We also must address node #3, which concerns the empirical magnitude of this increase in risk (which may be tempered by our confidence that our estimate is correct) and how small a risk we can justifiably ignore in the contexts in which the person is likely to operate; if this bias, again after discussion, is decided to be above that threshold, we then would need to fit it, according to the magnitude of risk, into a certain risk band, which would help to guide actions in light of that risk. Though the decisions at these three nodes are controversial, debate on harm, justice, and risk band is more likely to cut to the core of the controversy than a debate about whether this condition is natural or free from distinct sustaining causes.

OBJECTIONS TO REORIENTING AROUND RISK: PHENOMENOLOGY, BOATS, AND THE RELEVANCE OF PREDICTIVE BIOMARKERS

One concern, raised by Boenink (2009), is that the use of biomarkers and a focus on future risk of harm will cause us to lose sight of the patient's phenomenological experience, to discount the importance of what he or she is feeling at the moment. A similar concern is reviewed by Karlawish (2010), where he notes that some are concerned that the rise of risk assessment will erode the importance of the patient's chief concern. This need not be the case, however, if we are able to incorporate this phenomenology into our risk assessments, that is, that the presence or clustering of certain subjective symptoms can help us estimate the right risk banding more accurately than if we discard that information. Like Cordelia in *King Lear*, we can learn to balance the internal and external.

A second concern, which is voiced in Buchannan and colleagues' influential book, *From Chance to Choice*, is that better knowledge of our risks will prevent us from feeling in "the same boat," or will erode feelings of solidarity: "our ignorance of the pattern of distribution of deleterious genes has put us in a single lifeboat, each feeling vulnerable to disease. As the veil of ignorance lifts through genetic testing, those who are free of a given health risk are enabled to draw lines between the vulnerable and the secure" (Buchanan et al. 2000). Once we shift away from thinking in binary, however, the opposite outcome becomes likely. Given the complex landscape of biomarkers, where the risks are many and often reciprocal, we may merely have *differently concentrated* risks, with everyone at *some level* of risk; gaining knowledge of *some degree* of susceptibility to harms associated with disorders that these same people would otherwise shun may actually draw communities closer.

A third challenge would be from those who would accept the reoriented concept of disorder, but deny the relevance of predictive biomarkers to that concept. This style of argument may posit that there is something fundamentally different between bioprediction and the estimation of future harms based on traditional clinical signs and patient symptoms; an example is displayed by the opening paragraph of the preface to Jeremy Greene's "Prescribing by Numbers":

> This week [. . .] thousands of perfectly healthy-feeling adults will receive a diagnosis for a disorder that they did not know they had. There are several such disorders, imperceptible to patients, that produce no fevers, no chills, no headaches, no stomachaches, no pains. Neither are they immediately perceptible to physicians or to other health providers; there is no lesion to be seen with an ophthalmoscope or suspicious sound to be heard with a stethoscope, no tell-tale skin finding or sense to be made from piecing together disparate observations into a cohesive diagnosis. These are diseases that bear no immediate relation to symptoms but rather are connected to a statistical likelihood of developing symptoms in the future, pathologies—such as high blood pressure, mild diabetes, or elevated cholesterol—that are measurable only with the aid of an intervening diagnostic technology. (Greene 2007, vii)

Even classical clinical signs, however, require "an intervening diagnostic technology," such as the stethoscope that Greene mentions. The fever requires estimate via a thermometer. These classical clinical signs also have probabilistic relations to the harmful symptoms. Fever, quantified as temperature, is at the same time one of the most classic symptoms of disorder and a biomarker that is only probabilistically related to the appearance of a more general harm in the future.[21] If I detect fever, I might have a high probability of low-level discomfort over time, aches, and chills, and a low probability of a more significant harm such as death; I can better estimate those other harms by seeking other clinical signs.

The difference between the lesion, the suspicious sound, and the skin finding, on the one hand, and the high blood pressure, the low insulin response, or elevated cholesterol, on the other hand, is a historical difference, not one of kind. All are biomarkers; all contain probabilistic information about future harm. The difference is that the probability is quantified in the latter, but not the former case.

As I showed in the Introduction to this book, disorders defined by biomarkers are already commonplace (also see Table 4.2).

A CONCEPT OF DISORDER THAT IS ROBUST TO DEVELOPMENTS IN BIOPREDICTION

In this chapter I have proposed the Probability Dysfunction model of disorder: a condition is a disorder if and only if it is a variation from a biological image average that, after subtracting the results of injustice, is associated with an increased probability of harm in the future or present, and a probability function that justifies a certain response.

This model remains agnostic with regard to natural design, opting instead for variation from the image average. Unidirectionality is achieved by linking that variation to increased probability of future harm. I have noted that increased probability of future harm is the very core, albeit neglected, of the concept of disorder, and the shape of that probability as a function of time is capable of clarifying a number of long-standing debates. I have shown that probability functions of individual harms can be summated to establish disorder as a landscape that is capable of making sense of pleiotropy, and that this function can be used to establish specific disorders in terms of risk banding that matches risk to response. Finally, I have clarified nodes at which there is likely to be controversy remaining: (1) what counts as harm, (2) where to set the risk bands and whether there is threshold below which something ceases to be a disorder, and (3) whether an increase in risk of harm is due to injustice. In Table 4.4, we can see some examples of how this concept differs from the other concepts discussed in this section.

Although many classical conditions work well enough with a binary concept of disorder, a risk-reoriented one is best suited to the complex conditions common in the developed world (Vickers et al. 2008). The Probability Dysfunction concept is able to work synergistically, not antagonistically, with progress in biomarkers and as such is a concept that is able to handle the transition from "bedside medicine" to "desktop medicine" (Karlawish 2010). This transition is already in progress, with actuarial tools integrating handfuls of probabilistic markers to estimate 10-year risks for myocardial infarction (heart attack) or bone fracture.

I have shown in this chapter that thinking probabilistically in terms of risk of future harm is not something new, but has always been at the core of medicine. The difference is that we are finally getting the tools to bring that core to the surface and to recognize its more complex cases.

These more complex cases of disorder require the clinician, patient, and others to directly grapple with probability in a way that no longer lends itself to simplification into a binary. As the field of risk literacy has shown us, patients and doctors currently have great difficulty thinking of health in probabilistic terms (discussed briefly in Huang & Baum 2011; Vickers et al. 2008). The shift

Table 4.4 PROBABILITY DYSFUNCTION CLASSIFICATION COMPARED TO OTHER THEORIES OF DISORDER

Candidate	**Probability Dysfunction**	***DSM-5***	***DSM-IV-TR***	**Gert & Culver**	**Wakefield**	**Boorse**
Alzheimer's dementia	Y	Y	Y	Y	*? (not a clear dysfunction)*	*? (not a clear dysfunction)*
Osteoporosis (high risk of fracture)	Y	N (no present harm)	Y	Y	*N (no present harm)*	*? (not a clear dysfunction)*
Contraception	N	N	N	N (not harmful)	N (dysfunction but not harmful)	Y (interferes with reproduction)
Abnormal strength	N	N	N	N	N	N
Hypertrophic cardiomyopathy	Y	N (no present harm)	Y	Y	N	N
PKU in newborn	Y	N (expectable)	N (expectable)	N (external sustaining cause)	N	Y?
Broken bone or expectable depression (PTSD?)	Y	N	N	Y	Y	Y
Increased likelihood of impulsive violence	Y (increased likelihood of harm due to incarceration for violence in general is not unjust)	N (no present harm)	? conflict with society	?	?	?
Skin pigment in discriminative environment	N (no increase in harm once we subtract effects of injustice)	N (conflict between individual and society)	N (conflict between individual and society)	N (external sustaining cause)	N (not natural dysfunction)	N (not natural dysfunction)
Framingham risk score above image average	Y	N (no present harm)	?	? sustaining cause?	?	?

to conceiving of disorders according to the Probability Dysfunction concept, therefore, will require that we change the way we educate physicians (emphasizing statistics, computation, and risk communication) as well as the general population.[22]

As the cadre of biomarkers develops, moreover, we would move further away from a model in which determination of the risk band of a given harm is based on clinical judgment toward one based on quantitative and actuarial tools but *what to do about risk* continues to be based on clinical judgment in dialogue with the patient. This is analogous to the transition in forensics from clinical to increasingly actuarial risk prediction of future violence, where clinical predictions are largely unstructured assessments based on the judgment of clinicians and actuarial ones are completely structured—indicating in advance which risk factors should be considered and how they should be integrated and weighed (see Monahan 2014). The difference in the Probability Dysfunction model is that the variables in the algorithm would be limited to biological variations that increase risk of a harm not attributable to injustice.

With this probability-based way of thinking about disorder, we are now equipped to consider the implications of bioprediction on practical ethical debates within diagnosis (Part I, Chapter 5), responsibility (Part II), and distributive justice (Part III).

NOTES

1. Accessed at www.dsm5.org/proposedrevision/Pages/ proposedrevision.aspx?rid=465.
2. The explicit introduction of justice to the concept of disorder is likely to be controversial but is essential to exclude biological variation such as skin color and harm due to discrimination. I show that appeals to natural function (Boorse 1975; Wakefield 2007) or distinct sustaining cause (Gert et al. 2006) are untenable and justice preferable.
3. Though we can build evidence for natural function by looking at conservation of genes, pathways, or systems across species, the usefulness breaks down the closer we get to systems present primarily in humans. This is especially problematic with neurological and psychiatric disorders.
4. The ends of chromosomal DNA in most human cells are flanked by non-coding regions called telomeres that shorten with every replication; after many replications, the telomeres eventually erode completely, leading to the loss of coding DNA and catastrophic chromosomal abnormalities. Though these cells expressed during development an enzyme, called telomerase, which rebuilds the telomeres, humans specifically suppress telomerase in most differentiated cells (Nussbaum et al. 2007, 467).
5. In a sense, it is performing its function less well than it could be or than it once did; notice that this will *increase the likelihood* that the computer will be unable to function in contexts it could previously function in, like a 6-hour flight.

6. Willful abstinence (preventing the achievement of the natural goal of reproduction) might be another example of a dysfunction by the natural teleological account.
7. I will say more about the type of harm later in this chapeter, but for now, the thin concept is sufficient.
8. If everyone has major tooth decay somewhere, this might appear on the image average as low-level tooth decay on each tooth, and thus low level tooth decay would not count as a variation or disorder. The concept of variation changes over time. When most people had similar brain-cell death due to consuming water from lead pipes, this would not count as variation; but as lead pipes began to be phased out, the brain-cell death would begin also to phase out of the image average and would become a candidate for disorder. This temporal plasticity, I think, allows the *specific* disorders to change in a way that well mimics historical application while maintaining an overarching concept of disorder.
9. This value is also accompanied by a quantified measure of standard variation within the control sample. To extend the heart rhythm example, if the standard had a lot of variation at a certain peak in the rhythm, the image average may appear less sharp at that peak.
10. If a variation is associated with harm but a protective biological variation is found that abolishes the association with harm, then the group with the protective biological variation might be considered a candidate for a new reference class. This opens up the possibility that reference classes may become to be defined on the basis of *sets of biomarkers rather than demographics.*
11. If someone moves from eating very healthily and exercising hours daily to an average diet and lifestyle, this would increase risk of harm *for that individual* but not compared to the image average; the condition, therefore would make that person worse off, but would not count as a disorder. Gert and colleagues make a similar point (Gert et al. 2006, 129–164).
12. The risk of cervical carcinoma is also dependent on the type of HPV, which are separated by DNA sequence into high oncogenic risk and low oncogenic risk. The high oncogenic risk HPV 16 is implicated in an estimated 60 percent of cervical carcinomas. Most HPV infections are cleared by the immune system after a period of months, but high oncogenic risk HPV infections have a longer average time to clearance (13 months vs. 8 months) (Kumar et al. 2010, 1018–9). HPV infection might be thought of as a disorder that increased the risk of cervical cancer, and the type and extent of that infection might be predictive biomarkers for cervical cancer.
13. This idea is similar to Kaplan-Meier curves, which represent survival over time of different groups (Petrie, Sabin 2009, 133). Kaplan-Meier curves, if we substitute survival for periods of time free from a different harm, may be helpful in creating the probability curves I have in mind.
14. The exclusion of discrimination is another goal, which will be discussed shortly.
15. Accessed at http://www.dsm5.org/proposedrevision/Pages/proposedrevision.aspx?rid=465.
16. To deny that the risk clause is integral to our concepts of disease, one would need to jump through hoops to define it away as some sort of "present harm/dysfunction"

(i.e., fear). Because these harms would only exist if the person knew about their risk, the existence of disorder would depend on self-knowledge. Operating in probabilistic terms avoids this strange conclusion.

17. We might expand this analysis to consider biomarkers of increased probability of a *benefit*. Note, also, that this concept could easily incorporate uncertainty in the risks themselves by introducing a sort of weighting by confidence.
18. These decisions should be heavily informed by the patient's subjective judgment of the peaks and valleys of their personal probability functions (see Huang & Baum 2011).
19. Wakefield might argue that depression symptoms after bereavement would not be a disorder but a "natural" response.
20. Though a thorough defense is beyond the scope of this chapter, it could be argued that increased risk of harm due to rational desires and beliefs will often be cases of injustice.
21. One might object that the fever is associated with present pathology. When we see the following definition of pathology in the *Merriam-Webster Medical Dictionary*, however, it should be clear: "the anatomic and physiological deviations from the normal that constitute disease or characterize a particular disease."
22. This call to quantification is reflected in Karlawish (2010) and Vickers et al. (2008).

5

The Practical Ethics of Predictive Markers in Diagnosis

Can Risk Banding Address the Ethical Controversy Surrounding "Psychosis Risk Syndrome" and "Preclinical Alzheimer's Disease"?

In Chapters 3 and 4, I redefined "disorder" as a type of bio-risk. The most direct ethical impact of any theory of disorder is that it allows us to approach normative debates about whether we should recognize particular disorders. Because of the tensions between bioprediction and the classical binary way of thinking of disorders, it is precisely here that the first ethical challenges of bioprediction arise, and here that the Probability Dysfunction concept can most directly meet that challenge.

Two such normative debates have arisen recently within psychiatry and neurology in response to the introduction of probabilistic predictive information into formerly binary diagnostic categories. The first debate surrounds the behavior-based prediction of future psychotic episodes, and the second the bioprediction of dementia-like symptoms.

Though this book is specifically concerned with bioprediction, I include this debate around psychosis for three reasons. First, the issues raised in this context by behavioral markers and biomarkers are conceptually similar. Second, the debate around high probability states of psychosis is one of the most developed of its kind in psychiatry. Third, there is a large body of research in the pipeline aiming to supplement this behavior-based prediction of psychosis with bioprediction (e.g., the North American Prodrome Longitudinal Study; see Chung & Cannon 2015).

In this section, I show how the probability function (generally) and risk banding (specifically) can both add clarity to the debates and provide a practical solution that avoids many of the ethical concerns raised by high-risk states.

As I outlined in the Introduction, these two cases are just the tip of the bioprediction iceberg. Biologically based risk estimation of bipolar disorder, epilepsy, depression, Parkinson's disease, and even antisocial behavior is just beneath the surface. Before these risk categories progress further from bench to bedside, however, we need to seriously discuss how to conceptualize them. For how we do so will have a profound impact on the well-being of patients and the success or failure of the biopredictive enterprise.

ACTUARIAL METHODS, ACTUAL DIFFICULTIES

A polarized debate erupted from the proposed inclusion of an "at risk mental state" or "psychosis risk syndrome" in the *DSM-5* (Nelson & Yung 2011). The proposal was inspired by results from over 17 years of research identifying a subgroup of help-seeking young people with attenuated psychotic symptoms that is 20–40 times more likely than the general population to develop full psychosis (20–40 percent of the subgroup convert to full psychosis within several years, whereas the general incidence of psychosis is 1 percent over a lifetime). Effectiveness of early intervention in this subgroup, to delay or prevent conversion, is less well validated, but recent evidence has provided cause for optimism (Nelson & Yung 2011). With a range of predictive biomarkers—such as rate of cortical thinning in structural brain imaging (Chung & Cannon 2015), low omega-3 fatty acid content in blood cell membranes (Amminger & McGorry 2012), and the microfluidic multiplexing reviewed in Chapter 2—soon to supplement behavioral criteria, risk estimation is likely to improve.

Meanwhile, the research world of Alzheimer's disease buzzes about a set of genetic, fluid, and imaging biomarkers that many hope will identify people with "preclinical Alzheimer's disease" or "prodromal Alzheimer's disease," some of which are already being used in clinical trials such as the "A4" study (Anti-Amyloid Treatment in Asymptomatic Alzheimer's). These biomarkers include rare genetic variation associated with early onset disease, common genetic variants associated with late onset disease (e.g., the presence of the APOE ε4 gene variant), decreased levels of a protein called Amyloid Beta and increased levels of phosphorylated tau in the cerebrospinal fluid, gray matter loss on structural brain images, altered brain metabolism as shown by FDG-PET imaging, and PET imaging technologies to noninvasively estimate the burden of amyloid plaques and tau tangles in the brain (Ewers et al. 2011; Jack & Holtzman 2013; Sperling et al. 2011).

CONCERNS ABOUT RISK OF HARM IN ESTABLISHING CATEGORIES OF HIGH RISK OF HARM

The normative debate has so far turned on the risk of harm from labeling to those who would fall within the risk categories. The following quotations, made several years ago but still current, illustrate this concern in the case of dementia:

> [. . .] what if large-scale testing reveals that 20% or 40% of the general population over age 65 years have abnormal plaques on their PET scans?[1] Will we be doing more harm than good by telling such individuals, as one expert suggested, "you are on the Alzheimer road"? [. . .] After all, it is not yet clear that all patients with abnormal brain amyloid deposits will go on to develop full-blown Alzheimer's Dementia. (Pies 2010, p. 2)

Pies's "what if" scenario seems, by current empirical estimates, not to be outlandish at all; nearly one-third of cognitively normal elderly people have a high enough burden of the hallmark histological markers to satisfy the criteria for Alzheimer's disease at the time of death and *nearly all* elderly individuals have features of one type of brain pathology or another at autopsy (see Jack & Holtzman 2013). Pies's misgivings seem to revolve around uncertainty about the use of non-deterministic diagnostics and the lack of advantage to the patients gained by inclusion in this probabilistically defined category. Sperling (2011) expresses similar concerns:

> Use of these biomarkers in the clinical setting is currently unwarranted because many individuals [. . .] may not develop the clinical features of [Alzheimer's Disease] in their lifetime. Inappropriate use of this information in this context could be associated with unwarranted concern [. . .]. (Sperling et al. 2011)

Nelson and Yung (2011, p. 129) have summarized similar concerns about the advantages and potential dangers of including a psychosis risk disorder: "[. . .] the need for early intervention versus the risk of false positives, stigma and discrimination, the use of antipsychotic medication in at-risk individuals [. . .]."

Much time has been spent debating the empirical uncertainty that may modify this risk of harm to patients. What is the actual conversion rate? Will those falling within this definition suffer stigma or discrimination? Will there be inappropriate use of antipsychotic medication in this group? Comparatively little time, however, has been spent debating the appropriateness of how the proposed disorders themselves are framed, and how this frame itself modifies the risk of harm.

I aim to show that many of the harm-based objections (personal and impersonal) raised by opponents of including these disorders of risk are valid, but only insofar as we force these conditions into a binary conceptualization of the disorders. Consequently, these objections can be largely avoided by using careful *risk banding*, the useful heuristic developed in the previous chapter for estimating grades of risk. Finally, I will show that possible harms might be even further minimized if we adopt a program that prefers the development and use of predictive biomarkers rather than predictive behavioral markers.

LESS A DEBATE ABOUT DISORDER AS RISK BAND

In a 2011 paper, Nelson and Yung adroitly analyzed the debate surrounding the inclusion of a psychosis risk syndrome in the *DSM-5* and argued that despite the sometimes raised voices, the various sides of the debate actually agree on several key points:

> [. . .]the fact that people who meet these criteria are symptomatic and in need of clinical care, that they have an increased risk of developing a psychotic disorder, that they need treatment for current problems and to prevent transition to psychotic disorder, and that there is potential for harm in including the Risk Syndrome in DSM-5. (Nelson & Yung 2011)

In the previous chapters, I have argued that the risk clause is essential to any concept of disorder and that a person in possession of a set of biomarkers or other variation associated with significantly increased risk of harm should usually be considered as having some sort of disorder. This particular high-risk group, by Nelson and Yung's summary, straightforwardly meets the criteria of disorder according to the probability dysfunction. There is variation that is associated with present harm and increased risk of harm (insofar as psychosis is harmful), such that a certain response is justified, and little of this increased probability of harm is due to injustice.

Within the preclinical Alzheimer's disease debate, though many individuals exhibit mild cognitive impairment, some do not experience a clear present harm. They do display variation (the biomarkers) that researchers agree is associated with an increased risk of dementia-like cognitive impairment in the future.

Much of the controversy, therefore, seems to resemble Wakefield's (2007) objection to the WHO lumping all types of depression together (discussed in Chapter 4); namely, these objections are grounded primarily on the inappropriate grouping of disorders that should be within different *risk bands*.

There are good reasons for using different names for different disorders. I will argue that the primary benefit of using terminology is heuristic, to elicit certain intuitions about the important qualities of the disorder; a few examples are magnitude of risk of harm, type of harm, mechanism of harm, and what type of response is usually warranted. When a disorder's name has existed, for better or worse, for an extended period of time, the intuitions it elicits often become entrenched, and if risk landscapes illuminated by sets of biomarkers are tied directly to an existing name, it often inherits those same elicited intuitions. If the intuition elicited is not well matched to the risk band of disorder (i.e., to the magnitude and shape of its risk curve), it can cause unwarranted distress, inappropriate treatment, unwarranted risk of stigmatization, and other harms. Of course, if we switch entirely to thinking of disorder in terms of probability landscapes, the quantitative focus makes naming less of an issue. If we are to use this theoretical framework to have more immediate and practical influence, however, we might emphasize that thinking carefully about terminology so that it can elicit the most accurate intuitions about magnitude of risk (i.e., careful risk banding) can be an effective practical strategy to minimize these harms.

Here I adopt this sort of consequentialist framework in the analysis of the range of names that have been variously applied to these two controversial risk categories in psychiatry and neurology. I group them into several conceptual models, and I identify a curious gap in the models under consideration. I propose that reframing this research precisely within that gap, a model common in preventive cardiology, may decrease the risk of the harms of recognizing the proposed disorder, while making the benefits of care and early intervention accessible. As advances in bioprediction establish broader and more precise risk information, this model should be used in place of those based on "early diagnosis" or "end state".

THE CURRENT MODELS FOR NAMING HIGH-RISK CATEGORIES

The category with increased probability of psychosis has been associated with many names, from "ultra high risk," "clinical high risk," "prodromal psychosis," "prepsychosis," "subthreshold psychosis," to "at risk mental state," "psychosis risk syndrome," "risk syndrome for first psychosis," and finally, "attenuated psychotic symptoms syndrome" (Nelson & Yung 2011). Similar terms have been used for the high-risk category related to Alzheimer's disease. The Preclinical Alzheimer's Disease working group reports considering "asymptomatic," "presymptomatic," "latent," "premanifest," and "preclinical" Alzheimer's disease (Sperling et al. 2011); "prodromal" has also been used.

We can group most of these proposed terms into three main models: the prodromal model, the end-state model, and the sub-threshold model. These have varying strengths and weaknesses, and each invokes certain intuitions about the conditions' relationship to the established categories of schizophrenia or

Alzheimer's disease. I will start by considering the model I consider most ill-suited.

PRODROMAL MODEL

This model, which includes prodromal psychosis, prepsychosis, early diagnosis of psychosis, prodromal Alzheimer's disease, and early diagnosis of Alzheimer's disease, as well as preclinical Alzheimer's disease,[2] invokes the intuition that members of the identified group are *actually in* the early phases of the disease (psychosis, dementia, etc.). Prodrome, according to *Mosby's Medical Dictionary*, is defined as either "an early sign of a developing condition or disease" or "the earliest phase of a developing condition or disease" (Anderson et al. 1998). Early diagnosis, the simplest member of this group, implies that the person actually has psychosis or Alzheimer's disease, but the disorder has been caught early on.[3] Narrowly thus interpreted, this concept is obviously problematic and misleading.

Couched within the prodromal model is the idea that the probability of a patient becoming *x* in *y* period of time is nearly 100 percent; in the prodromal psychosis case, this is equally implied—that the case *will* develop to full psychosis. The terms "Alzheimer's disease" and "psychosis" have their own, historically entrenched, evoked intuitions about what kind of probabilistic relationship they have with their harmful symptoms—recurrence of psychotic episodes or cognitive dysfunction—and thus where they fall on a risk-banding hierarchy. If the risk band guides appropriate response, then the prodromal model implies that the condition identified by the behavioral or biomarkers effectively *has the same relationship* with the harms, or should be thought of as the same risk band, as schizophrenia or Alzheimer's disease. The lesser risk band collapses into the greater.

When we consider the phrasing of most critics, we can see that the most consistent objection is that the group identified by these predicted markers *is not the same risk band as the end stage.* Pies (2011) worries that "it is not yet clear that all patients with abnormal brain amyloid deposits will go on to develop full-blown Alzheimer's Dementia," and, to repeat Sperling: "Use of these biomarkers in the clinical setting is currently unwarranted because many individuals [. . .] may not develop the clinical features of AD in their lifetime" (Sperling 2011).

Most of the objections are set up in this structure: Being A implies a certain set of likely harms and appropriate responses, but B is not necessarily A, so treating B as A may lead to harm.

This prodromal model lends itself to the idea that the biomarker or behavioral marker should be thought of as a binary diagnostic test for psychosis or for Alzheimer's disease, rather than a category of its own with a different probabilistic relation to similar harms. Only if one operates on this prodromal framework are concerns about the effects of treating false positives and false negatives *of psychosis* or *of dementia* warranted.[4] I suspect that Mattson et al. adopt this model when they say the following: "if a false positive diagnosis results in treatment, any harmful side effect is a direct infringement of the basic medical ethics principle of non-maleficence, summarized in the Latin phrase *primum non nocere* ('first, do not harm')" (Mattsson et al. 2010).

Since the cases identified, by the prodromal model, have the earliest phase of the end-state illness to which they are tethered (psychosis), this model promotes deterministic thinking, and biases one to believe that the same treatments and attitudes of the end state might follow. So, for a prodromal psychosis: heavy medication and stigmatization.

Because objections of high rates of false positives and inappropriate antipsychotic prescription are most appropriate for the prodromal model, it seems likely that this is really the conceptualization to which most commentators respond.

END-STATE MODEL

Ultra high risk, clinical high risk, at risk mental state, psychosis risk syndrome, and risk syndrome for first psychosis all belong to this category, which, instead of dichotomously trying to identify cases and non-cases, takes a step forward by merely making probabilistic claims on the conversion to full psychosis or dementia (i.e., elevation to the next risk band). Thus, it has the strength that the language of false positives *of psychosis* and *of dementia* is inappropriately applied to this model. Worries are still expressed about overtreatment (perhaps raised for their own sake, but also possibly underlying concerns voiced about false positives on this model), about stigmatization, and about whether the conversion rate is sufficient to warrant its recognition. Most of these concerns result from the explicit connection between the term and an end state, rather than the probabilistic nature of the subpopulation.

If risk is integral to all disorders, as I argued in Chapter 4, but the magnitude and types of risk differ widely, defining the risk state by one possible end state (out of several) implies a very close relationship. Even high-probability risk disorders like epilepsy (the lasting predisposition or risk of seizure) do not include "risk" in the label. Similarly, including "psychosis" tethers the disorder to an end state (itself a heuristic for a particular arrangement of risk bands), which brings with it the risk of treating the condition the same way (overtreatment), people responding in similarly negative ways (stigma), and of identity defined by the foretold end state. Probabilistic information itself does not have these complications, but once explicitly tied to the end state, it gains them. Imagine talking about immune deficiency as fatal-illness-risk disorder or of a scraped knee as infection-risk disorder.

SUB-THRESHOLD MODEL

Sub-threshold psychosis and attenuated psychosis syndrome (APS) fit within this conceptualization. This model is an improvement over the prodromal and end-state models because it does not carry the connotation of *inevitable* progression and thus is resistant to concerns about low transition and false positives. It also better mimics the construction of probability thresholds that is a key feature of risk banding. However, sub-threshold psychoses and APS are still overly defined by the psychotic end state, and thus its stigma and medications; they are too close to a miniature version of psychosis. An instructive analogy within memory impairment is the separation of sub-threshold dementia-like symptoms into "mild cognitive impairment" rather than something like "attenuated dementia."

TAKING HEART

Across this range of terminology, a conceptual approach quite common in another medical realm is conspicuously absent: a nested hierarchical model of disorder, common in cardiology. Borrowing from preventive cardiology minimizes the potential harms (false positives and negatives, stigma, determinism, inappropriate antipsychotics) of recognizing these disorders in psychiatry and neurology while at the same time providing a well-established, familiar, and comparatively non-threatening vocabulary to talk about risk. This is a good model of practical risk banding.

A nested hierarchical model of risk better acknowledges both the biological forks (that only some in the category develop dementia or psychosis, while others develop a different disorder or none at all) and the magnitude of the risk band in terms of relation to harm. In such a model there is room for a hierarchy of risk bands that are all probabilistically linked to future harms, while permitting the existence of some bands with less devastating present harms. Atherosclerosis and heart disease are examples, where angina (chest pain) might be thought of as the distressing symptom that at the same time suggests a higher risk state in relation to the harm of a cardiovascular event.

If we anticipate—reasonably—that ongoing research will continue to develop better and better predictors of psychosis, Alzheimer's disease, and so on, we are likely to find that being at risk for a disorder and having a (different) disorder based on risk become one and the same. Though confusing at first, this interrelationship can be easily modeled as a nested hierarchy of risk bands of a certain harm along a dimensional axis.

For example, I may find that increased atherosclerotic plaque burden identifies a group of individuals at increased risk for heart attack. I also find that those with higher blood pressure, cholesterol, and inflammatory markers tend to be at increased risk of developing heavy plaque burdens. There is no reason I should not say that having lots of atherosclerotic plaques puts you at risk for heart attack, but that lots of atherosclerotic plaques is also a disorder called atherosclerosis, with its own probability curve. I can say that having higher blood pressure puts you at risk for atherosclerosis and also can call this high blood pressure "hypertension," its own disorder. In this way we develop nested shells of conditions that put you at risk for disorder A (heart attack) while also being disorder B (atherosclerosis).

To say that atherosclerosis is presymptomatic (or early diagnosed) heart attack or that high blood pressure is presymptomatic (or early diagnosed) heart attack (or atherosclerosis) would be misleading and potentially damaging. It might encourage overmedication, stigmatization, discrimination, and other harmful outcomes. If this were the terminology being used, *I might be right to bar the disorder from the diagnostic handbook.*

However, this nested hierarchical model makes it readily apparent that more work is required to show that medication for a higher risk band (psychosis) is appropriate for the lower band (a patient with elevated risk of developing psychosis), just as heart disease medications are often inappropriate for one with hypertension or high cholesterol. The type of intervention also changes appropriately at the different stage: lifestyle and statins for hypertension, omega-3s and general care for this new *DSM* model. Thus it encourages an appropriate matching of medications based on risk of harm. The clear separation of the risk bands avoids many of the types of issues raised in the analogical cases in psychiatry and neurology.

This model is also more flexible and able to capture the fluid nature of this subgroup of patients; many go on to a range of outcomes outside psychosis (other psychiatric disorder or none at all) or dementia (mild cognitive impairment or no progression), just like hypertension predisposes to both stroke and heart attack.

In some ways, my model of careful risk banding is similar to the clinical staging model championed by Patrick McGorry, which is imported from cancer medicine and proposes to establish the staging of psychiatric disorder like the staging of cancer. McGorry has advocated for terms such as "entry syndrome" and "common risk syndrome" via his clinical staging model as the first stage.

While this model has the potential to avoid some of the pitfalls of the prodromal, end-state, and sub-threshold models, a cancer model unfortunately first implies a binary (you have cancer) before the risk band (the risk of complication and death of having this cancer) and will not help to distance a risk category from its end state in a way that allays concerns of stigma. For this reason, I prefer priming thinking about graded risks with a preventive cardiology model, which is publicly well known and less feared.

AN ADVANTAGE OF BIOMARKERS IN NAMING RISK BANDS IN BRAIN-BASED DISORDER

Because this group at high risk of conversion to first psychosis is defined by characteristics that researchers have operationalized, quite literally, as subclinical manifestations of the symptoms present in psychotic disorders, there is not a convenient name for the disorder akin to high blood pressure. Perhaps "attenuated psychotic syndrome," or something similar, is the best we can get through a descriptive model, analogous to describing an arrhythmia as attenuated heart attack syndrome.

Development and use of biomarkers, by naming according to the internal variation that is associated with increased risk of harm, may enable us to discuss risk as fruitfully in psychiatry and neurology as in cardiology. This option is open for the group at high risk of development of profound cognitive impairment. A great start toward risk banding is actually made in Sperling (2011), as the three stage names of preclinical Alzheimer's disease: (1) asymptomatic cerebral amyloidosis (defined by the presence of amyloid markers only), (2) asymptomatic cerebral amyloidosis and "downstream" neurodegeneration (defined by the presence amyloid markers and markers of neuronal injury), and (3) amyloidosis and neuronal injury and subtle cognitive/behavioral decline (defined by the presence of markers of amyloid and neuronal injury and a behavioral marker).

SUMMARY AND TRANSITION

In this chapter I have discussed two normative debates surrounding the inclusion of disorders based upon predictive behavioral or bio-markers. I have shown that many of the concerns raised by commentators can be addressed by the adoption of the Probability Dysfunction concept of disorder and careful consideration of risk banding. As predictive biomarkers in psychiatry and neurology expand, similar ethical challenges within our binary structures of diagnosis are likely to arise. I have at the same time illustrated that the novel language of biomarkers has the potential to provide us with words divorced from stigma and untethered from frightening end states, just as the language of blood pressure has done in cardiology. In this manner, advances in bioprediction can work synergistically with the Probability Dysfunction concept and the idea of risk banding developed in Chapter 4.

The development of bioprediction also raises ethical challenges outside diagnostic manuals. I devote the rest of this book to an initial consideration of the ethical implications in two less anticipated areas: moral responsibility and distributive justice. In regard to the first issue, the estimation of the likelihood of an event in the future is not only central to the concept of disorder, but also to the concept of moral responsibility. In many cases, especially in psychiatric and neurological disorders, an agent may have reduced control or awareness of actions that may harm others. Can an agent, through bioprediction, gain probabilistic predictive knowledge of the likelihood of these harmful actions while still in a state of mind to influence the causal link between that action and the risk of harm to others? If so, does the development of bioprediction expand the range of an agent's obligations to the rest of society, and in this way enhance an agent's responsibility?

The intersection of bioprediction with moral responsibility will be discussed next in Part II.

NOTES

1. These numbers agree with current estimates; see Jack & Holtzman (2013) and Sperling et al. (2011).
2. The working group (Sperling et al. 2011) claims "the term 'preclinical' was felt to best encompass this conceptual phase of the disease process but is not meant to imply that all individuals who have evidence of early [Alzheimer's disease] pathology will necessarily progress to clinical AD dementia." Although the term invokes that intuition less strongly than the others they considered, it does so much more strongly than models not considered.
3. Even if you have been cosmically determined to die in 4 years' time, it would still be problematic and misleading to call you a dead person. Of course, the phrase "you're a dead man" *is* used colloquially.
4. A "false positive" would be any instance when one incorrectly identifies a non-case as a case. A "false negative" is an instance when one incorrectly identifies a case as a non-case. These commentators seem to be viewing the risk category as a diagnostic test for psychosis rather than a separate condition, as they talk about false positives *of psychosis* rather than of false positives *of risk of psychosis*.

PART II

Bioprediction and Moral Responsibility

Part II, which contains Chapters 6 and 7, is devoted to the intersection of bioprediction and moral responsibility. In Chapter 6, I will consider how the development of bioprediction can enhance moral responsibility. In Chapter 7, I will discuss how the development can diminish legal responsibility.

6

Enhanced Responsibility

Foreseeability and New Obligations to Others

Let us revisit "Risky Driving," which has been altered for confidentiality, from a real case:[1]

> Sally, a 33-year-old female, was brought into the emergency room after being hit by a bicycle as she crossed the street. Sally had a few scrapes and bruises, but was otherwise unharmed. Having had an uncle pass away unexpectedly from a heart attack, and feeling pain in her chest, Sally was quite distressed about her heart. An echocardiogram (an ultrasound on the heart) was performed. The echo-technician was surprised to find that the walls of Sally's heart were profoundly thickened. When Sally learned that the presence of this biomarker, increased heart wall thickness, corresponds to an increased risk of sudden cardiac death, the first thing she did was to make arrangements to stop driving the van for her children's playgroup. When asked her about that decision, Sally explained that she felt "it would have been irresponsible to do otherwise."

While there was little that clinically could be done to minimize Sally's risk of sudden cardiac death, she was incredibly glad to learn of that risk because it allowed her to have more control over the possible harmful consequences of that risk on those she held dear. In addition, she explained her change of behavior as the fulfillment of a moral obligation—a moral obligation not to risk involuntarily causing harm.

Sally's response is interesting because, if she is correct, it suggests that the development of bioprediction may not only be clinically useful, but *morally useful.*

Sally's intuition is echoed by the results of the *REVEAL* study, which examined the feasibility and outcome of communicating risk status of Alzheimer's dementia (based on APOE ε4 status). In general, there was interest in learning one's status in order to plan for the future should one indeed become incapable of managing for oneself, and to prepare family, despite the absence of clinical treatment (Roberts & Tersegno 2010).

Was Sally right to postulate such a moral obligation? One way to examine this question is to consider an inversion. That is, if Sally had continued to drive the van and had experienced a cardiac event that led to a crash and the harming of the children, *should we hold her morally responsible*? In other words, if she failed to meet this putative obligation to avoid risking harm to others, would she have acted wrongly?

EXPANDED BIOPREDICTIVE TECHNOLOGIES, ENHANCED MORAL RESPONSIBILITY

In this section, I appeal to foreseeability-grounded theories of moral responsibility to answer the questions raised by Sally's intuitions in the affirmative. These theories, like Sally, posit that an agent can properly be held responsible for an act even when he or she exhibited no meaningful control or awareness during the act, as long as the possible harmful effects of entering a state of involuntariness or ignorance were foreseeable to the agent. I argue that the application of these foreseeability-grounded theories highlight hitherto under-discussed reasons to know one's own biological risk, that is, to enable control over negative effects of possible disorder on those within one's community.

I argue that the risk of the future adverse event underlies the moral weight of the diagnosis in terms of an obligation to protect others. In this way, the assignment of moral responsibility is similar to the definition of disorder developed in Part I; the difference is that the component of interest is not risk of harm to self but risk of harm to others.

In this discussion, it becomes readily apparent that not nearly enough time has been spent discussing the proper role of degrees of foreseeability. Like the classical concept of disorder, foreseeability gets us into trouble if we approach it in binary. With the development of bioprediction and the quantification of risk, we can no longer operate in terms of foreseeable/not foreseeable. How foreseeable do harmful outcomes of my possible future loss of control need to be for me to have such obligations? How strong are the obligations?

One standard we can look to goes back to the social obligation surrounding epilepsy and driving. Whatever the likelihood of a seizure while driving, we could treat consistently biopredictive technologies able to estimate the same magnitude of risk. Whatever that would be is the threshold of risk at which the obligation requires the cost of not driving. When we try to do so, however, I show that the standard is so high that we seem poised to shoulder an immense and crippling burden of obligations.

Using a discussion of culpable ignorance, I go on to discuss whether the obligations are not only to act on this risk information for the sake of others, but to gain the information, and whether this extends into secondary obligations to increase access to and/or develop bioprediction. Throughout this discussion, the demandingness of these obligations should become clear.

Finally, I propose three strategies for dealing with this expansion in the domain of moral responsibility: (1) the solution from cost, (2) the solution from risk banding, and (3) the solution from technology.

1. Since we operate under limited resources, responsibility should be limited by the cost and availability of biopredictive tests.
2. We should adopt *risk banding* in determining blameworthiness; the strength of the obligation (and thus the degree of blameworthiness for failing to meet the obligation) should be proportional to the magnitude of risk and type of harm.
3. We should devote resources and set priorities to move biopredictive technologies away from identifying permanently at-high-risk groups (simple genetic associations being the prime example) to identifying more temporally restricted risks.

I argue that the solution from cost is morally deficient. We should adopt a combination of the latter two, as they allow agents to meet the obligations to others without significantly impairing the ability to pursue a fulfilling life.

JUDGMENTS OF MORAL RESPONSIBILITY

There is much controversy about the concept of moral responsibility (e.g., see Eshleman 2009 or Fischer 1999 for a review). What follows in this chapter is necessarily a simplification, but I aim to review the concepts most relevant to bioprediction.

What we mean when we say that someone is morally responsible for something is that one of his actions makes him the appropriate target of a set of reactive attitudes like praise or blame (Fischer 1999; Strawson 2008; Wallace 1994). The act of holding responsible, explains Wallace, reflects a belief that a moral obligation has been violated.[2]

The modern views that an agent can *truly be* morally responsible, and thus can be an *appropriate target* of reactive attitudes, descend from Aristotle's *Nicomachean Ethics* (Aristotle 1999, Book III). Integral to Aristotle's concept is the requirement that the agent's action be voluntary in order for the agent to properly be held morally responsible. For Aristotle, an action is involuntary if it is compelled or performed in ignorance.[3]

Most modern theorists have interpreted Aristotle's voluntariness as two requirements for moral responsibility: (1) a control condition (control over whether to carry out the action), and (2) an epistemic condition (knowledge of what one is doing) (Aristotle 1999). Moral philosophers thus differentiate between the wrongness of an action and an agent's blameworthiness for that action (see Smith 1983). If one violates a moral obligation without knowing what one does, or knows what one does without having any control over whether to do it, one cannot properly be held morally responsible (or judged to be blameworthy). These are examples of excuses. Many acute psychiatric or neurological conditions—seizure, delusion, unconsciousness, cognitive impairment—have the potential to violate either the control or epistemic conditions.

Much of the debate on moral responsibility flows from this Aristotelian concept, and the competing views can be roughly described as variants of two main categories: (1) the merit-based view, and (2) the consequentialist view.[4]

In the merit-based view, blame would be appropriate only if the agent merits it. Watson argues that an action must first be attributable to an agent, that is, an action must *truly belong* to an agent for him to merit blame (Watson 1996). In the accountability view of moral responsibility, to hold an agent morally responsible is to require the agent to give an account, an excuse or justification, for her behavior or to accept that she is properly the target of reactive attitudes like blame (Oshana 1997).

In the consequentialist view, blame would be appropriate only if blaming would likely lead to a beneficial change in the agent's behavior or character (i.e., the agent's ability to control the action could be influenced by blaming). While most debate over the past several decades has been about variants of the merit-based view, the consequentialist view is slowly regaining traction, largely because it is viewed by some as more resistant to the threat of determinism (see Eshleman 2009).

In an influential essay, "Freedom and Resentment," Strawson (2008) argues it is precisely that we feel an agent is the appropriate target of this type of moral response, which he terms "reactive attitudes," that *makes* an agent responsible. In other words, an agent is morally responsible simply if society in fact applies to him or her a set of reactive attitudes. This argument is not useful in the evaluation of bioprediction, however, as the attribution of reactive attitudes in conjunction with novel technology is likely to be erratic.

Wallace, however, argues that an agent is morally responsible if it is *fair* that he or she is a target of reactive attitudes, which is a judgment of the (non)expression of (good) ill will in the agent's action (Wallace 1994). This Strawsonian account (like the accountability view) has the advantage of allowing the critique of *existing practices of holding responsible* that may not be appropriate judgments about cases that do not yet exist.

FORESEEABILITY IN MORAL RESPONSIBILITY

Foreseeability (and thus bioprediction) is most relevant to the epistemic condition of Aristotle's requirement. In order to know what I am doing, I need to have an idea of what consequences my action is likely to bring about.

If I throw a rock, I need to foresee that it is likely that my action may result in the breaking of your window in order for you to appropriately blame me for this action.[5] In other words, foreseeability provides a reason to do otherwise when alternate possibilities of action are available to me.[6] It creates a basis from which I can form intentions. From a consequentialist perspective, foreseeability (of the consequence of my action or merely *that* I will be held responsible) is necessary for any sort of behavioral change.

Some theorists (Glannon 1998; Zimmerman 1986) argue that foreseeability of likely consequences allows us to hold someone responsible for a state of affairs even through the person had no meaningful control over his or her actions, or did not know what he or she was doing, at the time the state of affairs comes about.[7]

Glannon argues that foreseeability enables "remote causal control," that is, "the cognitive ability to respond to theoretical reasoning concerning the foreseeable consequences of our actions and omissions that are responsive to practical reasoning" (Glannon 1998). Glannon distinguishes this type of remote causal control from the "proximate causal control," at the time of a particular action, and argues that only remote causal control is necessary for moral responsibility. Responsibility for actions during the period of non-control shifts to that earlier time at which a decision was made with remote causal control.

Zimmerman, too, embraces this foreseeability-grounded account of responsibility when he develops an account of negligence.

> . . . [S]omething may be out of one's control at the time that one does it (or fails to do it) and yet not be out of one's control entirely—and the fact that it is not out of one's control entirely provides the opportunity for one's being properly held morally responsible for doing it (or failing to do it). We may say: although my causing the damage was not in my immediate control, it was in my remote control. It is just this sort of picture that I wish to draw of negligence. (Zimmerman 1986, p. 206)

As illustration of these theories, consider the case of Reckless:

Case 1 (Reckless): Reckless is driving a truck down Main Street. Suddenly, she begins to speed up and swerve through traffic until her truck goes out of control and veers wildly off the road onto a sidewalk, where the truck impacts, and severely injures, two pedestrians.

Once the truck spun out of control, Reckless could no longer influence whether she harmed the pedestrians (she no longer had proximate causal control). However, we do not view this lack of control at the time of impact as an excuse for harming the pedestrians. Because the vehicle crash and harming of pedestrians was a foreseeable consequence of Reckless's speeding and swerving, and thus in her remote causal control, we can appropriately hold Reckless responsible.

Glannon summarizes his position as follows in regard to mental disorder:

> A person may voluntarily perform some action that causes a disordered mental state during which wrongful acts are committed, and yet be responsible for these acts because of the voluntariness of his earlier act and because he was capable of foreseeing [and taking steps to avoid] the likely consequences of his earlier action, specifically the wrongful acts committed in the disordered state. In such a case, responsibility transfers from the earlier act to the later consequences, and the individual in question would be responsible both for the action and the consequences. (Glannon 1998, p. 245)

We can contrast the case of Reckless with the case of Serendipity.

Case 2 (Serendipity): Serendipity is driving a truck down Main Street. Suddenly, Serendipity suffers a spontaneous seizure. She loses consciousness and her truck veers wildly off the road onto a sidewalk, where the truck impacts, and severely injures, two pedestrians.

The consequences of Reckless and Serendipity's traffic accidents are the same, the type of driving behavior is the same, and it may be that the very movements of the drivers' limbs could have been the same. Once we find out about Serendipity's seizure, however, our moral assessment turns on its head.

At first glance it seems obvious that we should not blame Serendipity—after all, a seizure is often used as the definition of involuntary action. Although it is important to keep legal responsibility and moral responsibility separate, it is worth pointing out that the law in most jurisdictions would excuse

Serendipity. Though the consequences were grave, she would not be held responsible for the injuries of the two pedestrians because the involuntariness of her actions negates the required voluntariness requirement.[8] We should not say that Serendipity ought to have behaved otherwise if she indeed had no ability to do so (Coggon 2008; Goddu 2006).[9]

At second glance, however, we will notice that once Reckless spun out of control, her action was also involuntary. Neither driver had proximate control at the time of impact. What is different is that while Reckless had remote control over the crash, Serendipity did not. Serendipity had no reason to foresee losing control of the truck due to having a seizure.

Still, it is not entirely true that Serendipity lacked remote causal control. Before the seizure onset, she did possess "the cognitive ability to respond to theoretical reasoning concerning the foreseeable consequences of our actions and omissions that are responsive to practical reasoning" (Glannon 1998, p. 233). Had she foreseen that a seizure was likely, she would have been able to act on that knowledge (by pulling over) to avoid harming the pedestrians.

In order to distinguish between Reckless and Serendipity, therefore, we could follow Zimmerman (1986) to distinguish between "standard control" and "enhanced control." My being aware of the possibility of a certain outcome enables enhanced control. Because Reckless was aware of the possibility of the loss of control after speeding and swerving, she has the requisite *enhanced* remote control of the sort Glannon seems to envision for "responsibility to transfer from the earlier act to the later consequences" (Glannon 1998, p. 245). Serendipity does not have enhanced control, so she is appropriately excused from blame.

Zimmerman and Glannon are not the first to make this kind of argument for the centrality of foreseeability in moral responsibility; Aristotle made a similar argument in *Nicomachean Ethics* (Aristotle 1999). In Book III, Aristotle intimates that a person may be held morally responsible for actions performed while intoxicated with alcohol if he voluntarily drinks at an earlier time and thereby risks acting wrongfully, albeit in ignorance, at a later time. Glannon cites this Aristotelian example, in fact, in support of his own theory (Glannon 1998). Zimmerman provides a similar thought experiment about wittingly taking a pill with the foreseen consequence of "running amok" at a later time and causing damage (Zimmerman 1986). In these cases, all the agents had enhanced remote control over the future action.

When we consider the case of Foreseeable, however, we may begin to realize that these foreseeability-grounded theories of moral responsibility require further specification:

Case 3 (Foreseeable): Foreseeable is driving a truck down Main Street. Suddenly, Foreseeable experiences a seizure. She loses consciousness and her truck veers wildly off the road onto a sidewalk, where the truck impacts, and severely injures, two pedestrians. Foreseeable had been previously diagnosed with epilepsy and had not taken her anti-epileptic medication that morning.

Glannon uses a very similar case as paradigmatic of the transfer of responsibility to an earlier action:

> Foreseeability and the transfer of responsibility suggest that [. . . if someone with schizophrenia] voluntarily fails to take prescribed antipsychotic medication to control his deluded beliefs and is able to foresee the likely consequences of failing to do so, then he may be responsible for what he does during a psychotic phase [. . .] even though by themselves these acts are performed on the basis of disordered beliefs or volitions which would appear to constitute excusing conditions. (Glannon 1998, pp. 242–243)

As we saw with Serendipity, a seizure would usually excuse an individual. But with Foreseeable, the reasoning goes, she could foresee the likely consequences of her voluntary decision to drive without having taken the anti-epileptic drugs, that is, having a seizure while driving, which could lead to a loss of control and the possibility of harming those nearby. She had the relevant enhanced remote control over the harm. In the absence of an *independent excuse* for why Foreseeable decided to drive after not taking the anti-epileptic medication that morning, she is blamed.

Another way to answer why we blame Foreseeable but not Serendipity might be to use the terms of negligence; because the seizure was foreseeable, and because there were proper precautions that could have been taken but were not, the involuntariness of the seizure can no longer negate moral responsibility. As a result of this negligence, she should be held morally responsible for the pedestrians' injuries, even though the direct effects of her seizure were as involuntary as Serendipity's. This response, too, is grounded in foreseeing the likelihood of future seizure. We feel that the moral responsibility is anchored by Foreseeable's earlier choice to drive the car even though she hadn't taken her anti-epileptic medication, an action over which she did have control.

BIOPREDICTION AND FORESEEABLE PROBABILITIES

Let us take stock of the moral judgments we have established (and in these cases, the legal structures would mirror these judgments):

1. Reckless should be held morally responsible (enhanced remote control: foreseen probability of loss of vehicle control).
2. Serendipity should not be held morally responsible (no enhanced remote control: spontaneous seizure).
3. Foreseeable should be held morally responsible (enhanced remote control: foreseen probability of seizure).

The case of epilepsy and driving was purposefully chosen, as it is a condition in regard to which we have relatively well-developed social intuitions and policies that implicitly weigh foreseeable probabilities and proper precautions. In what follows, I will argue that the moral justification of these obligations ultimately reduces to the probability of seizure causing harm to others; probability information gleaned from bioprediction, therefore, is similar in terms of moral relevance and should be treated consistently.

Now let us consider a fourth case, in which we finally test Sally's intuition:

Case 4 (Probability): Probability is driving a truck down Main Street. Suddenly, Probability experiences a seizure. She loses consciousness and her truck veers wildly off the road onto a sidewalk, where the truck impacts, and severely injures, two pedestrians. Probability had never before had a seizure, but Probability had previously gained access to biopredictive testing, which informed her that she was at high risk for seizure within that year.[10]

Probability appears to have had the same enhanced remote causal control as Foreseeability. Probability voluntarily made a decision to drive with the foreseeable consequence of having a seizure while driving and risking harm to pedestrians (see Appendix II for a short review of the science of seizure prediction).[11]

The consequences are the same, the seizure is the same and equally as involuntary, and the voluntary choice in both cases was to drive with the foreseeable risk of seizure. Foreseeable had not taken anti-epileptic medication that morning, but Serendipity did not take medication either.[12] Nonetheless, there is likely to be resistance to conceding that we should blame Foreseeable and Probability to the same degree; one might object that the diagnosis is more serious than the biopredictive tests.

In this section, I will show that Foreseeable and Probability's cases are equivalent in all morally important ways, and therefore we should treat them consistently. I will address what I view to be the strongest objection remaining to blaming them equally, which I call the objection from diagnosis.

The Objection from Diagnosis

> *The Objection from Diagnosis*: it is morally important that Foreseeable had a diagnosis of epilepsy while Probability had only a probabilistic risk.

One might claim that there is an obvious difference between these cases; Foreseeable has a diagnosis of epilepsy and Serendipity does not. This difference justifies treating them differently. Since the risk of seizure is held constant between the cases, however, the objection from diagnosis must posit that the diagnosis is morally relevant for some reason above the identification of a risk of future seizure.

Our first investigation might be into whether a diagnosis *per se* is what does the moral work. It is immediately obvious that not any diagnosis will do; if Foreseeable had a diagnosis of schizophrenia and had not been taking her antipsychotic medication when she suffered a seizure while driving the truck, we would not hold her morally responsible. It is, rather, that the diagnosis of epilepsy, specifically, says something meaningful about the involuntary behavior *in which we are interested* (i.e., the seizure).[13]

Is there something morally relevant to the diagnosis of epilepsy specifically? The most recent and authoritative definition of epilepsy proposed by the International League Against Epilepsy (ILAE) and the International Bureau for Epilepsy (IBE) is the following:

> Epilepsy is a disorder of the brain characterized by an enduring predisposition to generate epileptic seizures and by the neurobiologic, cognitive, psychological, and social consequences of this condition. The definition of epilepsy requires the occurrence of at least one epileptic seizure [. . .] An epileptic seizure is a transient occurrence of signs and/or symptoms due to abnormal excessive or synchronous neuronal activity in the brain. (Fisher et al. 2005, p. 471)

From this definition, we learn that in having the diagnosis of epilepsy, (1) Foreseeable has previously had a seizure, and (2) Foreseeable is likely to experience a future seizure. This latter component is the probability component, but what about the former? Might having previously had a seizure be morally important in a different way?

The relevance of former seizure at first seems fruitful, as Probability has never had a seizure, and many jurisdictions actually do prohibit driving in the period 6 months to 12 months following a seizure (Devlin et al. 2012). But this obligation not to drive exists without the need for the diagnosis of epilepsy. The justification for the obligation is that the experience of a past seizure provides information on the risk of future seizure.[14]

It might be clear by now that the objection from diagnosis is no more than a manifestation of the reification of existing categories of disorder that I highlighted in Chapter 3. I argued in Chapter 4, moreover, that a disorder is primarily a statement of the type and magnitude of risk of harm to self. The value of a diagnosis of epilepsy is precisely a statement about magnitude of risk of future seizure. The past experience of seizure, the EEG brain wave patterns (a predictive biomarker), and other elements of the diagnosis are valuable in this context precisely because they help us estimate the magnitude of risk. In this case, that risk of harm to self (seizure) translates through the vehicle to a risk of harm to others. Whether the magnitude and type of risk are identified through classical diagnostic pathways or novel means should not matter morally.[15]

If we establish through other means that someone has the same probability of future seizure, we should treat her as we did Foreseeable. Conversely, if we could show that someone who had previously had a seizure, or who has a classical diagnosis of epilepsy, is no more likely than the general population to have a seizure in a vehicle in the future, then we should treat her like Serendipity, that is, excused if indeed she suffers a seizure in the vehicle.[16]

Like Serendipity, Probability had neither a past seizure nor diagnosis. But like Foreseeable, the likelihood for Probability to have a seizure was elevated compared to the general population. Moreover, both Foreseeable and Probability (but not Serendipity) *knew* about this increased risk and voluntarily chose to drive, thus risking the seizure and the harm to others. Probability could also have taken precautions against the seizure (decided not to drive), but also decided not to do so that day. So the seizure was foreseeable, and precautions could have been taken against it. In other words, in both cases the agents exhibited enhanced remote causal control. In this moral framework, we should hold Probability and Foreseeable morally responsible for the pedestrians' injuries.

If everything collapses into probabilities, we should treat like cases alike and consider the novel case of Probability similarly to our more familiar case of Foreseeable. Such a conclusion involves recognizing that bioprediction expands the domain of moral responsibility.

How Foreseeable Should a Consequence Be for Moral Responsibility?

Glannon's foreseeability-grounded theory of moral responsibility provided a good platform from which to start thinking, but it assumes that it is obvious whether consequences of my actions are foreseeable or not foreseeable. Bioprediction, by allowing the quantification of a range of specific risks, forces us to realize that a dichotomy is not at all obvious.

What we are left with are questions about the type of foreseeability (is it what I do foresee, what I could foresee but did not, or what I should foresee but did not?). We could also ask about the concreteness of the wrongdoing that is foreseen (I might foresee a possibility of harm to myself due to seizure, but not to others), the role of probability (does it matter how likely the wrongdoing is to occur? i.e., I might have actual foresight of a very unlikely but possible state of affairs), and what determines the appropriateness of precautions that should be taken to meet the obligations to others (what restrictions to my freedom can I be expected to make?). While these issues are not new (Montmarquet 1999; Nelkin 2011; Sher 2006, 2009; Smith 1983), bioprediction forces us to be more precise in our answers.

Up until this point, for example, we have been assuming that the probability of seizure leading to accident (and thus harm to others) is the same in both the case of Foreseeability and Probability. But it should become pretty clear that it is not at all obvious what kind of probability we are assuming, or exactly how it relates to moral responsibility. How high an estimated probability is necessary for me to "foresee" a consequence in a way that is relevant for moral responsibility?

Even Reckless, who seemed to be the most straightforwardly blameworthy, did not foresee with *certainty* that dodging and weaving through traffic would cause her to lose control of the truck. Similarly, it may be that if Foreseeable remained un-medicated, her likelihood of having a seizure *sometime in the future* approached 100 percent, but the likelihood of her having a seizure *while driving the truck on that particular trip* could still have been quite low.[17]

Zimmerman defends a definition of negligence that at least comments on the degree of probability. He maintains that P is to some degree negligent if P foresees that there is a "non-zero probability" that a harmful/wrongful event will occur in the future unless P takes a possible precautionary measure, but unjustifiably decided not to do so, and the event indeed happens (Zimmerman 1986).

This threshold of foreseeability, too, seems unhelpful. The development of biopredictive technologies (and the quantification of risk in general) virtually guarantees that we will all have "non-zero" probabilities. With a single "-omic" assay, you will instantly have thousands. Even without the assay, we should know that there is a small but certainly "non-zero" risk that any given person will have an unprovoked seizure and that this may occur while that person is driving. Maybe there is a threshold of risk under which I am justified in neglecting small risks?

On this question, however, Zimmerman is unhelpful: "It is not my purpose here to try to say just what makes a decision unjustifiable, and so I shall not be concerned with how it is that one may tell that a case is a case of negligence rather than of justifiable neglect. That is a separate, and very difficult, issue." So while he acknowledges that magnitude of probability, seriousness, and ease of precaution of the foreseen harm/wrong likely weigh into whether an agent's decision in regard to precautions are unjustified, he is silent on the cases where I foresee only the *chance of running amok*, which is precisely the type of case with which we are faced.

One way to make sense of the world of foreseen *probabilities* of harms (not foreseen harms) is to take a page from Strawson's book. If we accept the uneasy solution that moral responsibility is, at least in part, a function of existing social practices of holding responsible, then we can look for a relevantly similar case.

One possibility is using the driving restrictions set on people with epilepsy, which we have been using as a marker of threshold at which we hold responsible; unfortunately, however, it is anything but clear from the literature what probability of harm this group risks to others by driving. Estimates, based on patchy data from around the world, place relative risk of accident anywhere from one to three times that of the general population (see Devlin 2012). Moreover, risk varies widely within this group, with some epilepsies easily controlled with medication, and others having seizures mostly during sleep. What emerges from an analysis of how people with epilepsy are held responsible is that there are strict sanctions for an under-defined risk.

Another possibility is to look at policies surrounding obligations of people who have experienced a single unprovoked seizure (Bonnett et al. 2010). The most informative was developed by the UK Driving and Vehicle Licensing Agency in 2009, and permits those who experience a first seizure to regain their private license after a period of 6 months, as long as no biomarkers (e.g., EEG epileptiform activity) indicate that their 12-month risk of seizure is greater than 20 percent (Bonnett et al. 2010). Within the context of this book, it should be clear that whatever threshold is established, the time from last seizure (and the experience of last seizure) should matter only insofar as it contributes meaningfully to the risk estimate. That is, people with a 12-month risk of seizure less than 20 percent should not be held responsible. This threshold is acknowledged as arbitrary, but it gives us a starting point in our considerations of if and at what level such a threshold should be set. I will return to a discussion of risk-thresholds for foreseeability later in this chapter.

When we take that 20 percent 12-month risk as the minimum foreseeable probability of harm to others to be held responsible for a decision to make a single trip, what do we get? If, to simplify quick calculation in the absence of the data set, we assume that the 20 percent risk is evenly distributed throughout the hours of that 12-month period (which it almost certainly is not), and assume driving for 10 hours a week (an hour to work and home for 10 trips per week), then we get a (very) rough per trip risk of seizure while driving to be ~0.04 percent (or 1 in 2,500).[18] If we use this threshold for responsibility, and

I voluntarily drive to Blenheim Palace with the foreseen probability of seizure of, say 0.1 percent, then I would be held morally responsible if, during that trip, I have a seizure that causes me to lose control of the car and injure the Duke of Marlborough.

Putting aside the question of whether I can really be said to foresee, in any meaningful way, an event that is predicted not to happen 999 times in 1000, I might wonder whether a *per-trip* calculation of risk was appropriate. Perhaps Sally's intuition of responsibility was based on exposing others to *cumulative* risk of harm?

If cumulative exposure is important, a much lesser probability of per-trip harm spread over a great time could lead to a higher incidence. It is clear that some concept of the cumulative exposure is built into current driving restrictions concerning people with epilepsy as the seizure-free period that needs to be demonstrated in many districts before one can reapply for a driver's license jumps significantly if one wants to drive as an occupation. If blame can come in degrees (after we decide to hold responsible), this would suggest that Probability, who was driving every day of the year without any extra precautions, might be seen as more blameworthy than Foreseeable, who drove only that one day without precautions.

Since driving itself is risky, however, extending this line of reasoning renders it no longer clear how we could justify treating differently a full-time taxi driver, who drives 8 hours a day, from the person with epilepsy who drives a few days a year and thus has the same cumulative yearly risk of causing an accident. Such cumulative scores would be especially demanding when considering career choice. However, it is highly counterintuitive that there should be obligations on professional drivers to drive only a certain amount per year or over their career *because they are professional drivers.*

While this statement may seem like a *reductio ad absurdum*, it might instead hint at a more general obligation to take increasing precautions when one frequently engages in a potentially harmful activity.[19] A taxi driver might have an increased responsibility for driving carefully, and a professional medical researcher might have an increased responsibility to respect confidentiality. Just what kind of responsibilities the general quantification of risk creates is worthy of further discussion, but it is beyond the scope of this book.

Because when we talk about moral responsibility, we usually talk in terms of specific acts and their consequences, I will move forward with the per-trip model of foreseeability.[20] Since I have shown that the moral obligation to avoid risking harm to others is the same whether identified through bioprediction, historical diagnosis, or past event that is predictive of future event, and that this 20 percent 12-month risk already exists, we should examine its logical extension.

In the previous sections of this chapter, I have applied foreseeability-grounded theories of moral responsibility to a series of cases to show how prior knowledge of the probability of future harm to others is grounds for an agent to be held responsible for harmful actions, even when these actions would otherwise be fully excused. I showed that bioprediction is capable of revealing the right kind of information about possible future risks to others. I have shown that the degree of foreseeability must be specified, and have proposed a tentative practical solution by linking the necessary degree of foreseeability to the existing UK policy on risk of seizure deemed acceptable to return to driving.

Both Glannon and Zimmerman's foreseeability-grounded accounts provide useful starting points but leave us with many questions as we strive to extend and critique the role of foreseeability in responsibility. The development of biopredictive technologies forces us to consider the roles of subjective and objective foreseeability, the concreteness of the possible consequences, and the role of the probability, seriousness, and ease of precaution against the possible consequence.

The main issue to which I will now turn is whether the availability of this type of bioprediction creates an obligation to find out one's own risks in order to meet the previously established moral obligations to the others in one's community.

BENIGHTING ACTS AND CULPABLE IGNORANCE

The previous section briefly raised the question of how concrete foresight must be to bring moral responsibility. Zimmerman (1986) suggests that the foreseeability is not what is logically, physically, or reasonably possible to foresee; rather, it is a subjective foreseeability, and so responsibility can extend only to events that are *actually foreseen (to some degree)*. However, this has the counterintuitive result that you might be able to reduce your responsibility by intentionally limiting your knowledge of risk. Some element of hypothetical or ideal foreseeability must be included in an adequate foreseeability criterion, but what, exactly, *should* one foresee, even if one did not? Is there such a thing as foresight that one *should have had*?

Consider the case of Blissful:

Case 5 (Blissful): Blissful is driving a truck down Main Street. Suddenly, Blissful suffers a seizure. She loses consciousness and her truck veers wildly off the road onto a sidewalk, where the truck impacts, and severely injures, two pedestrians. Blissful had never before had a seizure. Blissful had previously knowingly forgone the opportunity for free, safe, and confidential biopredictive testing,[21] which would have informed her that she was at high risk for seizure within that year.

Blissful's situation differs from Probability's solely by Blissful's decision not to get tested. Because of this decision, she did not have the information that may have guided her action toward taking precautions against unwittingly harming the pedestrians. She lacks the same sort of enhanced control upon which we reasoned that Probability was morally responsible for the harm caused to the pedestrians. Nonetheless, I argue that we should also hold Blissful responsible.

Let me address several objections to this step in the argument.

On the one hand, Blissful did not know that she was putting others at risk in the same way that Probability knew, and therefore seems to lack the enhanced remote control necessary for moral responsibility. On the other hand, Blissful did know that bioprediction would enable her to find out her risk. Because she chose not to find out, one could argue that the enhanced remote control shifts further backward in time to the earlier decision not to be tested. She could foresee (to some degree) that her willful ignorance could put others at avoidable risk.

The term for a decision not to gain this kind of knowledge, taken from the literature on culpable ignorance, is a "benighting act."

> "Benighting Act": an initial act, in which the agent fails to improve (or positively impairs) his cognitive position, [followed by] a subsequent act in which he does wrong because of his resulting ignorance. (Smith 1983, p. 547)

The decision not to take the biopredictive test is a failure to improve her cognitive position in precisely a way that creates a risk of unwittingly risking harm to the pedestrians, and thus fits the criteria for the benighting act. If the commission of such a benighting act renders someone culpable for her ignorance, it might follow that we hold Blissful morally responsible, along with Probability and Foreseeable.

Now, one might argue that accepting this benighting standard would mean that it is morally required for everyone to take this biopredictive test, if the risks and benefits were similar. While this is indeed a possibility, a thorough explication is beyond the scope of the present discussion. Rather, it aims only to raise the issue.

In making this leap, one might reasonably ask if it matters how likely it is for someone to be in this high-risk category. If the answer is one in 100,000, or one in 100, one might argue that this might differentially affect our moral assessment. However, acknowledging such would merely show the moral obligation to be *defeasible* by practical concerns such as how readily available and affordable the test is. Our initial moral assessment would not change; it would merely mean that we might have to balance it with practical considerations. If the test is freely available, safe, and confidential, as in Blissful's case, the proportion of people who fall into the high-risk category should not matter. If I were to run over your dog when I reversed in my car, it would be no excuse to say that I chose not to perform the predictive test of looking in my rearview mirror (once the mirror exists) because I did not know anyone on the block had a dog (and so, thought the number of people at high risk for suffering a run-over-dog was negligible). The ease of gaining information about whether there is something behind the car impacts the judgment of whether acting in ignorance of that information is culpable, even when the likelihood of a dog is very low.

The field of culpable ignorance offers three major theories that may help us in our analysis: the Conservative, the Moderate, and the Liberal, to borrow Smith's (1983) terminology. Conservative theorists, Aristotle being one, posit that culpable ignorance provides no excuse for wrongful behavior; they would hold Blissful, Probability, and Foreseeable fully responsible for the harm to the pedestrians. Moderates posit that culpable ignorance provides a partial excuse; they would hold Blissful, Probability, and Foreseeable responsible for the harm, but would argue that Blissful is *less blameworthy*. For liberals, the agent's ignorance, however culpable, *fully excuses* him for performing the *unwitting wrongful act*. He is held to blame, however, for performing the earlier *benighting act*. For the liberal, a benighting act that is highly likely to result in an unwitting act is more blameworthy than one that is less likely to do so; its blameworthiness also varies as a function of the turpitude of the unwitting act.

For the purposes of this chapter, it is not necessary to accept the Conservative thesis that Blissful, Probability, and Foreseeable are blameworthy to the same degree, but only the weaker thesis (Moderate or even Liberal) that Blissful is *more blameworthy* for risking harm to pedestrians *than is Serendipity*.

Accepting even this weaker thesis would mean that the availability of such a biopredictive technology creates a *pro tanto* other-regarding obligation to use it.[22]

Such other-regarding obligations to obtain biopredictive services have hitherto been under-addressed in the literature on predictive testing, which has tended to focus heavily on reasons related to medical treatment. For example, the American Academy of Pediatrics and the American College of Medical Genetics recommends against any sort of predictive genetic testing in asymptomatic individuals under the age of 18 unless a beneficial clinical intervention exists (Pal et al. 2010). While the existence of a beneficial clinical intervention is certainly an important factor in such a recommendation, the existence of a beneficial moral intervention should also be considered. Even if no effective clinical intervention exists, knowing that you are at high risk for seizure would enable people to show a compassionate consideration for the potential impact of their actions within the community, by taking special precautions or avoiding risking harm to others.[23] Although the American Academy of Pediatrics and the American College of Medical Genetics may indeed have considered such moral usefulness in their recommendation, it is important to be sure to do so.

BIOPREDICTION DOWN THE RABBIT HOLE

If we give credence to the weak thesis that Blissful is at least more blameworthy than Serendipity, we have one more case to consider:

Case 6 (Uninformed): Uninformed is driving a truck down Main Street. Suddenly, Uninformed suffers a seizure. She loses consciousness and her truck veers wildly off the road onto a sidewalk, where the truck impacts, and severely injures, two pedestrians. Uninformed had never before had a seizure. Uninformed did not know that biopredictive testing for seizure risk exists. Had she known, and had she undergone the testing, it would have informed her that she was at high risk for seizure within that year.

Could we argue that uninformed *should have known* about this test? Could we argue even the weaker position that uninformed is also, in some way, more blameworthy than Serendipity? From Uninformed's perspective, *she is Serendipity.*

But *we* know better. From the perspective of the scientists, doctors, academics, critics, and policymakers who know of the existence of the bioprediction, Uninformed's seizure could have been foreseeable. Could we argue, therefore, that *this knowledge of the existence of the test* is somehow relevant to whether we hold Uninformed responsible? It certainly seems the case that there is something importantly different between Serendipity's unpredictable seizure and this one.

Again, this case raises the problem of foreseeability criteria, and how that standard should be determined. Should it be subjective foreseeability (what the agent foresees), objective foreseeability (what could be foreseen), or some hybrid hypothetical foreseeability (what the agent would have known had he or she followed certain steps)? From an objective standpoint, Uninformed may have acted wrongly, but from a subjective standpoint, she did not.[24] All the theories presented thus far in this chapter, including Smith's development of culpable ignorance, posit subjective foreseeability; they require some prior voluntary culpable action or omission with some amount of subjective foreseeability of risk of future harm/wrongdoing.

Interestingly, a 2008 cognitive science experiment suggested that objective foreseeability is important in folk morality (Lagnado & Channon 2008). The researchers had participants rate moral scenarios and found that both subjective foreseeability (what the depicted agent foresees) and objective foreseeability (what the third party knows to be the case) significantly impacted decisions to blame, regardless of whether the action was intentional or unintentional. It is contested how much we should take from such an experiment, however.

Sher (2009, 88) attempted to create a hybrid model in which an agent could be held morally responsible even if he is unaware that the act is wrong "despite having evidence for its wrongness or foolishness, his failure to recognize which a) falls below some applicable standard, and b) is caused by the interaction of some combination of his constitutive attitudes, dispositions, and traits."

The applicable standard is an adaptation of the legal reasonable man standard: what a reasonable man would have done in the same situation. The failure to do as the reasonable man would do needs to be caused by the fixed psychological features of the agent. There are several issues with this standard (Nelkin 2011). For example, Sher considers physical features of the agent, such as poor eyesight, to be part of the reasonable man, but psychological features, like unusual moral evaluations, not to be part of the reasonable man, and thus runs afoul with a physical/psychological binary distinction which I address in Chapter 7. As such, the relevance to Uninformed is not at all clear.

Unless Uninformed's ignorance of the existence of the biopredictive test can be shown to be due to a prior wrongful benighting act, therefore, it seems we should tentatively conclude that moral responsibility ends at Blissful.

It does seem, however, that the possession of the knowledge of a biopredictive test by a certain class of person (doctors, academics, scientists, etc.) raises questions about their role in relation to this particular risk of harm. Perhaps part of the reason is the intuition that objective foreseeability reflects an obligation not on Uninformed's part, but on ours. It raises the possibility of an other-regarding meta-obligation to help others gain the enhanced remote control necessary to discharge their own other-regarding obligations. Here, this might manifest in a *pro tanto* obligation to raise awareness of such a biopredictive test, and an obligation to disseminate the products of biopredictive research. This touches more broadly upon the obligations of experts to apply that knowledge in public and community life. Taking another step further, might we have now struck upon an obligation to *develop* biopredictive technologies?

NOT JUST SEIZURE

At this point, I think it is worth noting the sheer extent of the domain over which moral responsibility seems poised to expand with the development of biopredictive technologies. We are beginning to predict risk not only for seizure, but for aggression, schizophrenia, dementia, and the list is fast growing.

Aggression

In Chapter 7, I discuss the body of work that suggests that boys exposed to an adverse environment during childhood (childhood maltreatment) develop an increased predisposition toward episodes of impulsive aggression only if they also possess a common, low-expressing version of the monoamine oxidase A (MAOA) gene. Boys with the other common MAOA variant who were also exposed to an adverse environment during childhood are just as likely to express impulsive aggression as boys of both genotypes who were not exposed to the adverse environment. I argue in Chapter 7 that under certain conditions such an interaction might reasonably reduce criminal responsibility, but this may be true only if, like the seizure case, the predisposition is unknown at the time of the act. Once this bioprediction exists, are boys who fall into the high-risk category actually again responsible for their actions if they do not take proper precautions against the manifestation of this predisposition?

Psychosis

In Chapter 5, I introduced that it is already possible to identify a group of children 20–40 percent of whom will go on to develop a first psychotic episode and the debate about inclusion in the next revision of *DSM*. The precision of the prediction is likely to improve along with new biomarkers. Currently, an unpredicted relevant first psychotic episode may serve as a complete excuse in criminal court, but will it still be available for this high-risk group?

Dementia

Also in Chapter 5, I introduced the progress of predictive biomarkers for dementia symptoms. MRI volumetric measurements of the hippocampus and entorhinal cortex (integral in consolidation of memories and the first place of plaque and tangle formation in Alzheimer's disease), PET imaging of plaques, and cerebrospinal fluid proteomics assays for amyloid beta are all vigorously being pursued as predictive markers of dementia. Many of the same issues of driving and epilepsy also apply to dementia, as does acting in any role where proper cognitive function is necessary to protect others from reasonable harms (surgeons, airplane pilots, child caretakers, etc.).

If we treat all these new biopredictions like we treat epilepsy, where a *one in 2500 per act risk of harm to others* is the threshold of obligations strong enough to prohibit driving, it seems like our obligations to take proper precautions to protect the public from our future selves is about to explode. This is especially true when we begin to imagine the actuarial combinations of individual smaller risks enabled by bioprediction.

How might we avoid this bio-risk responsibility expansion problem?

A TENTATIVE ATTEMPT AT A SOLUTION

How might we deal with this explosion of other-regarding obligations? We might still be able to work within the new framework to define limits for what we are in fact responsible. I will present and briefly expand upon three possible solutions: (1) the solution from cost; (2) the solution from risk banding; (3) the solution from technology.

The Solution from Cost

Most would agree that we are not obligated to do *everything* to make others in our communities safer, especially if doing so carries significant costs. The solution from cost is perhaps the most straightforward solution, therefore: the idea that ultimately the costs of these biopredictive technologies and of the proper precautions must counterbalance their claims on us. The relevant costs could be monetary (i.e., the price of the test and precautionary measure), but could also be thought of in terms of time, degree of infringement on autonomy, and so on. In the seizure prediction case, for example, if the biopredictive tests were very expensive (due to new technology), inconvenient (e.g., require 48-hour hospitalization), or invasive (e.g., an electrode to the dentate gyrus), we might not feel that Blissful behaved wrongly when she passed up the opportunity to take the test.

It is important to note, however, that the practical considerations merely indicate that the moral obligation created by the bioprediction is defeasible, that the solution from cost is actually an independent excuse for violating the obligation. It is for this reason that the solution from cost may be acceptable (and perhaps inevitable), but unfortunate. The solution from cost is essentially an acknowledgment that a barrier prevents individuals from discharging their moral obligations in the best way. It is an acknowledgment that states of affairs would be much better if the barrier to access were removed from the technology; if we were in a position of governmental power, therefore, we might see this set of events and strive to improve the accessibility of these technologies.

What level of cost is acceptable to demand? If we use the obligations of people with seizures and epilepsy as an example (and remember that the actual risk this category poses when driving has not been precisely quantified), the cost required of individuals to take precautions seems high. For example, common anti-epileptics cost on average \$200-\$1200 in out-of-pocket fees per year in the USA depending on insurance status (Halpern et al. 2011) and can have significant side effects. The driving restrictions on those who have experienced a first seizure (nearly half a year of no-drive time) and those with epilepsy (5–10 years demonstrably seizure free) are high, and apply even if one had previously been dependent on driving for employment or alternative methods of transportation are unavailable. In New York City, not being able to drive might not be a big deal, but in rural Montana, it is effectively house arrest. Note also that the restrictions are not altered in areas of low population density, where the risk of harming others if one were to have a seizure in a car is significantly diminished. Under this high bar, a large amount of cost would seem supported indeed.

Full elaboration of the costs we might reasonably require of people to improve (through taking a biopredictive test) one's baseline knowledge of the risks one might pose to others is beyond the scope of this chapter and should be a focus of future work in the neuroethics of biomarkers. As the multiplexing of biomarkers and automation of their analysis becomes more routine, however, it may fast become as cheap or cheaper to assay huge sets of biomarkers than the single-gene or single-metabolite tests common to routine medical care; we are already seeing this happen on a small scale when a physician is interested primarily in a single assay but orders a panel because the lab no longer offers (or charges more for) the individual tests. If this trend continues, then an increasingly large fraction of cases will resemble Blissful, where the costs of bioprediction are very low. If we accept the conclusions from our discussion of Blissful, moreover, then we may be able to adopt a very simple calculus in a large number of cases.

As these technologies and the computational tools to link them together proliferate, therefore, we will inevitably have to find a method to manage these risks with limited resources in a way that does not leave our normal lives shackled. One practical solution would be to agree upon a priority list of risks for which bioprediction and proper precautions would be required. This list very well might have a different appearance depending on the social role one is to take on (bus driver or pilot, for example, as opposed to a teacher). Creating such a list in a way that is just and with minimal stigmatization, however, is likely to be an immensely difficult enterprise. In the practical list solution, moreover, what would happen to the risks that have not made the list? Simply, members of society might agree to tolerate certain risks of harm to each other.

The Solution from Risk Banding

Another possibility might be the position that we have hitherto been far too harsh on the small categories of people, like those with epilepsy, of whose risk we have historically been aware. The demandingness of bioprediction on moral responsibility, therefore, may be avoided by resetting the threshold. The result might be an elevation of the threshold of acceptable risk of harm to others, and a softening of the stringency of what we determine to be proper precautions against that risk. An objection, however, is that we should not "go soft on risk"—that the obligations entailed by those risks should, if at all possible, be met.

The alternative, therefore, takes a page right out of the concept of careful risk banding developed in Chapter 4. Rather than a threshold of responsibility, we should have stratifications based on the magnitude and type of foreseeable risk to others. We can import the Probability Dysfunction concept and combine it with a foreseeability-grounded theory of moral responsibility that, like the moderate (or liberal) theories of culpable ignorance, allows a more readily graded (and thus personalizable) scale of obligation and blameworthiness.

The Solution from Technology

There remains a third possible solution that could feasibly address weaknesses in both the solution from cost and the solution from risk banding. The problem with the current biopredictive technologies is that they establish categories of *sustained elevated risk*, that is, elevated risk categories that span large lengths of time. In this way, they are very similar to the diagnostic category of epilepsy and its definitional component of sustained predisposition toward seizures. Constructed this way, everyone in these categories becomes a "ticking time bomb" that could explode at any moment without warning. Thus, the people in these categories seem to be required to take proper precautions throughout the entirety of their risk window. Empirical evidence suggests, however, that the events these conditions estimate are *not* equally likely to occur at all times in the risk window. Since the events we are trying to predict are themselves discrete and periodic, if we could better home in on indicators of risk that rise in a smaller time window around the event, we might be able to live relatively unencumbered lives while meeting the obligation not to expose others to an unreasonable risk of harm.

To use seizure again as an example, there are groups developing EEG-based technology combined with machine learning that has the potential of identifying brain activity that is highly likely to spill over into seizure within a short interval (Williamson et al. 2012). With this technology (perhaps embedded into a baseball cap that could plug into a car's dashboard), one could reasonably be driving a car and pull over for an hour if the risk of seizure goes above a certain threshold value defined by traffic safety. This would enable even those with frequent seizures to gain the benefits of driving and the mobile life without putting the public at risk. The practical extension of this solution, therefore, would be to make this kind of short-time-window bioprediction a research priority, rather than funding merely biomarkers of nebulous temporal frame.

MOVING RESPONSIBLY FORWARD

In this chapter, I have investigated the intersection of bioprediction with foreseeability-grounded theories of moral responsibility. I argued that biopredictive technologies can be *morally useful*, enabling us to fulfill a domain of other-regarding obligations the violation of which we would otherwise have no control over. Anchoring bioprediction of risk of harm to others to our existing practices toward people with epilepsy, however, I showed to be a suboptimal solution. The domain of other-regarding obligations enabled by biopredictive technologies is likely to be too demanding for individuals identified with risks that surpass that threshold, so I considered three possible solutions to this enhanced domain of responsibility. After rejecting the solution from cost as morally deficient, I endorse instead a combination of the application of risk banding to responsibility and a focused change in the type of bioprediction that we develop most aggressively.

It may be counterintuitive that, having pointed out the special historical relation of brain-based disorders with stigma, blame, and discrimination in Chapter 2, I devote this time to argue that bioprediction—especially of brain-based risks—enhances the domain of actions/omissions for which we can be held responsible. However, it is particularly because of this history that I am quick to highlight the problem, so that we, foreseeing the risk of exacerbating the history, might fulfill our obligation to consider the problem early and deeply.

As we develop bioprediction, we will necessarily pass through a time when we will encounter many cases like Uninformed, where the technology exists, but where that existence is known only to a handful. For certain disorders, that is the current state of affairs. As long as the pathways that elevate risk of harm to others are not as clear as the risk of loss of vehicular control due to a seizure, moreover, we must be aware of a proclivity to restrict too strongly the freedom of those who fall within the easily comprehended pathways. In Chapter 7, I consider arguments that suggest how the development of bioprediction should in certain cases, like that of Uninformed, mitigate legal responsibility. Satisfactorily approaching this discussion, too, will involve shifting our thinking from binary to probability.

NOTES

1. I have written briefly on a similar case, along with my colleagues, in Huang et al. (2014).
2. Moral responsibility is inherently social, and thus differs from other common language uses of the word "responsible" (Eshleman 2009). One might say, for example, that the Colorado River is responsible for carving the Grand Canyon.

We acknowledge the causal connection between the river and the creation of the canyon without feeling that we should praise or blame the river.

3. There is an extensive literature on free will and moral responsibility (Fischer 1999); I acknowledge this debate, and that I cannot engage with it sufficiently here. The challenge of free will for moral responsibility is typically framed in terms of the control requirement, but bioprediction interfaces most directly with the epistemic condition.
4. Consequentialists may object that there really is not a consequentialist view on moral responsibility, merely that a set of responses to actions that resemble the practices of moral responsibility can sometimes be justified on consequentialist grounds.
5. For the early parts of this chapter it is sufficient to operate on the simplistic view that appropriateness of blame is influenced by foreseeability. I will elaborate finer distinctions in regard to that influence in later parts of this chapter. Many think that if an agent actually did foresee a potential consequence and acted with that knowledge, for example, this is worse than if the person should have foreseen but did not and acted in ignorance; this difference is sometimes referred to as recklessness versus negligence.
6. A colleague recently brought to my attention another view, one of the proponents of which is Nomy Arpaly, that foreseeability provides a metric by which we might judge the degree of ill will embedded in an action.
7. These foreseeability-grounded theories are similar to the "Tracing" view of epistemic control discussed in Sher (2009), as reviewed in Nelkin (2011).
8. We might think that Serendipity should feel remorseful, but we would not hold her morally responsible; see the distinction between the involuntary (action done in ignorance, which the agent regrets) and the non-voluntary (action done in ignorance, which the agent does not regret) in Aristotle (1999, Book III.1).
9. The literature surrounding "moral luck" addresses situations where we seem to hold an agent morally responsible even though she does not have control, but will not be visited here.
10. The precise probability is not important here. Let us assume that the risks of seizure in the car are nearly equivalent or greater on a per-trip basis in Probability's case than in Foreseeable's.
11. As we can see in Appendix II, the real potential in these technologies lies in their combination.
12. Since anti-epileptic medications are often depressants, their removal might remove inhibition of the brain's electrical circuits. While disinhibition would increase the likelihood of seizure, such an increase theoretically could be matched by a requisite risk predicted by the biopredictive test.
13. If foreseeable had had a diagnosis of schizophrenia, had not been taking her antipsychotic medication, and had crashed into the pedestrians as a direct result of a delusion or hallucination (i.e., the behavior that the diagnosis predicts and the medications take precautions against), then we might hold her responsible. The point is that the behavior that is predicted is the cause of the accident.
14. There may be value in the experience of a first seizure if this causes the agent to take the risk more seriously now that he knows what it is like. This raises the

difficult topic of how to decide how seriously we should take risks of particular magnitudes; I briefly consider this topic at the end of the chapter.

15. This assumes we are reasonably confident in the accuracy of each estimate.
16. Australia is considering increasing the use of biomarkers to enable those with epilepsy at low risk of seizure to return to driving sooner (Devlin et al. 2012).
17. Recently, Levy suggested that epistemic luck completely undermines moral responsibility (Levy 2011). I do not hope to justify, here, my resistance to Levy's conclusion, and my claims remain open to his objection.
18. Although there are a great many types of seizures, not all of which would lead to a vehicular accident, for simplicity I will assume loss of consciousness.
19. Some might argue that it matters whether there will be less driving total if the person drives less, or if another driver will just drive more hours; if we care primarily about cumulative risk to drivers and pedestrians in a society, we should increase obligations proportionally to the likelihood this will decrease cumulative risk (i.e., only in situations where the driver will not be replaced). Others might argue that the degree to which cumulative risk is concentrated within a given individual matters because it reflects the risk that the individual is willing to impose on others, or in our case, the probability that the *particular individual* foresees that she herself will bring about a certain state of affairs.
20. One possible justification for this move is to consider what a cumulative model would mean for a car salesman. If the car salesman sells a large number of cars over his lifetime, the foreseeable risk of serious harm to others due to a car accident will approach 1, but it would seem absurd to blame a car salesman. However, the per-act risk would be quite low.
21. Imagine that seizure risk became part of a clinical multiplex panel assay and thus running the assay for one biomarker or 1000 costs virtually the same.
22. There are good consequentialist reasons to hold Blissful responsible. If we were to hold a higher standard of responsibility for those who take the test and test in the high-risk category than for those who don't take the test at all, those wishing to avoid responsibility would actively seek ignorance. This would be akin to drivers who drive without their headlights to avoid responsibility for collisions they did not see (because they were driving in willed ignorance). A similarly questionable approach is to suggest that researchers performing whole genome analysis can avoid a moral dilemma on whether to inform the research participant of an incidental finding by intentionally not analyzing certain areas of the genome, such as the APOE ε4 locus.
23. Many would like to know their risk of Alzheimer's disease even though there is no adequate treatment. Knowing risk can allow better planning and deliberation of end-of-life care and minimize the negative impact of such an outcome on family and loved ones.
24. See Jackson (1991) for an account of how the objective rightness of an action is dependent on the subjective: "We need, if you like, a story from the inside of an agent to be part of any theory which is properly a theory in ethics, and having the best consequences is a story from the outside."

7

Reduced Responsibility

Distinguishing Conditions in Which Biomarkers Properly Reduce Legal Responsibility

INTRODUCTION

In the previous chapter I suggested that if an agent is aware of a risk of harm to others that falls beneath a threshold of foreseeability, or if she is non-culpably ignorant of a risk, then she may not be blameworthy if harm occurs. This chapter considers the extension of these criteria to more clearly articulate the conditions in which biomarkers might be exculpatory in the context of legal responsibility.

I begin the argument by responding to general objections to the use of probabilistic information (often gleaned from groups) in judgments of legal responsibility. By appealing to the same sort of argument upon which my Probability Dysfunction model of disorder was developed in Chapter 4, I argue that these general objections are incoherent. I then suggest that one place that neural biomarkers may become relevant to criminal law is through the "reasonable man" standard, which is used in some jurisdictions in conjunction with the partial defense of provocation. I engage with the objection that the reasonable man precludes the relevance of a biological predisposition toward violent behavior; I argue that in cases where there is sufficient evidence that the likelihood of violent response is influenced by *certain legally relevant mechanisms*, such as increased emotional gravity of the provocation to the person, that biomarker evidence is probative to the legal issue. I frame my arguments in the context of three real-world cases in which testing was performed for a predictive biomarker, the monoamine oxidase A (MAOA) gene-environment interaction, and an argument made for reduced responsibility.

THE ROAD TO REDUCED RESPONSIBILITY

Consider the following case of Candice Anderson, whose story about a probabilistic predisposition to a car crash was reported in the *New York Times* on November 25, 2014:

> *Candice Anderson received the bittersweet news Monday in a Texas courtroom, fighting back tears, and her arm around the mother of the boyfriend she had felt responsible for killing in a car crash 10 years ago.*
>
> *The judge cleared Ms. Anderson in the death of the boyfriend [Mr. Erickson] even though she had pleaded guilty to criminally negligent homicide in the case years ago.*
>
> *Ms. Anderson, 21 at the time of the crash, was driving her car when she inexplicably lost control and crashed into a tree. Mr. Erickson, her passenger, died at the scene, and Ms. Anderson has been racked with guilt ever since.*
>
> *In getting her record cleared, Ms. Anderson benefited from an extraordinary—and long delayed—admission by General Motors, which on Monday for the first time publicly linked Mr. Erickson's death to an ignition switch defect in millions of its small cars …*
>
> *Ms. Anderson's Saturn Ion was among the cars equipped by G.M. with the defective switch, which can cause a loss of power, disabling power brakes, power steering and airbags. At least 35 deaths have been link to the defect, which went unreported by G.M. for more than a decade …*
>
> *Until this year, she wrestled with questions about her role in Mr. Erickson's death. The police trooper who investigated the accident had deduced that Ms. Anderson was intoxicated before her drug test results came back. His police report referred to the seemingly inexplicable circumstances of the accident, her history of recreational drug use, "and Anderson's behavior at the scene," which was disoriented and emotional …*
>
> *The district attorney who prosecuted Ms. Anderson […] and the police trooper who investigated the accident had both said that if the ignition-switch defect had been publicly known at the time of the crash, certain details of the accident—like the lack of skid marks or evasive action—would have been seen differently …*
>
> *[The DA said] "Had I known at the time that G.M. knew of these issues and has since admitted to such, I do not believe the grand jury would have indicted her for intoxication manslaughter." [… A senator x called the first indictment] "a perversion of the justice process," [… and Ms. Anderson said one benefit of the situation would be that one day she would be able to tell her kids … this is what happened] "and it wasn't mommy's fault." (Business Section pp. A1 & B2, Nov 25, 2014)*

This is a complicated case, and we do not have all of the relevant details, but I have included it because it raises interesting propositions about the intersection of probability and reduced legal responsibility that are of relevance to the neuroethics of biomarkers.

These propositions include:

A. The probabilistic predisposition to shutting off the power and steering is legally relevant in some way, even though the occurrence of that event would not "guarantee" a violent crash.
B. The knowledge of that risk in court would validly have changed how other evidence "was read."
C. This change in how other evidence would have been "read" validly would have led to a reduction in legal responsibility.
D. That G.M. is blameworthy for holding but not disclosing this information about the predispositions of others.

When translated from the realm of mechanics to biomarkers and from probabilities of steering control to self-control, each of these propositions is likely to be controversial (for this reason I ground the discussion in this more familiar case of cars). I will consider each of the propositions A—C in turn, through analogy to the oft-cited case of boys who possess a genetic variation that only when combined with an adverse childhood environment increases the likelihood of future impulsive violent behavior, i.e. the monoamine oxidase A (MAOA) gene x environment interaction (first reported in Caspi 2002, summarized in Baum 2011). I will return to proposition D later in this book when I discuss institutional responsibilities in light of the development of biomarkers.

I have previously written on the research on MAOA (Baum 2011), so I will only briefly describe it here. Humans predominantly possess one of two variations in length of the MAOA gene, and because these different variations show different levels of activity of the product of that gene (a protein that breaks down certain classes of neurotransmitters) in in vitro assays, these two common genetic variations are often referred to as MAOA-H (high in vitro activity) and MAOA-L (low in vitro activity). A now reasonably replicated result confirmed by several meta-analyses is that there is a higher risk of future impulsive violence in boys (at least boys of European decent) who were exposed to an adverse childhood environment (e.g., abuse) *and* possess MAOA-L, while boys with MAOA-H are resilient to abuse by this measure. To set the magnitude of the effect in context, the authors of the first study (Caspi et al. 2002) noted that 44 percent of the violent convictions were attributable to the small fraction (12 percent) of observed boys who had a combination of MAOA-L and childhood maltreatment; the authors further noted that this effect on risk of violent conviction was of similar magnitude by one measure (attributable risk fraction) to the effect of the major cardiovascular risk factors on the risk of cardiovascular disease.

The MAOA gene-environment interaction can be intuitively understood with a simple car analogy. Imagine that a car manufacturer has two factories, each that uses a similar but slightly different metal contact in its ignition switch system. Cars from one factory have "metal-H" and those from the other have "metal-L," but they are otherwise indistinguishable. Post-market surveillance reveals that cars with "metal L" are more likely to have car accidents than cars with "metal-H," *but only if the car habitually had been operated in a very rainy environment* (e.g., Seattle). Thus the effect of metal-L on increased crash risk occurs only when combined with an adverse environment. In dry environments, it doesn't matter at all which metal your car has.

There are many caveats and limitations to existing research on the MAOA gene-environment interaction (see Baum 2011; Baum & Savulescu 2014; Buckholtz et al. 2014), including questions about what counts as an adverse environment, the time window of exposure to the environment to have the effect, and the populations to which the results generalize (for example, the case seems much more complex for females). However, because the MAOA gene-environment interaction is comparatively well supported in the field of biomarkers of antisocial behavior and has been invoked in several real-world murder trials in order to reduce sentence or responsibility of a defendant, it has become a useful discursive tinderbox for the neuroethics of biomarkers.

The three cases are those of *Bayout* (2009), *Waldroup* (2009), and *Albertani* (2011). I have reviewed them in more detail elsewhere (Baum 2011; Baum & Savulescu 2014), so I will summarize them briefly.

In *Bayout* (2009), the accused slit the throat of a young man whom the accused had mistakenly taken to be one of a group of young people who had assaulted the accused earlier in the day (Ahuja 2009; *Bayout v. Francesco* 2009; Feresin 2009; Forzano et al. 2010). At a first trial, the accused had been found guilty but had been given a reduced sentence of 9 years on hypothesis of some reduction of rational capacity as evidenced by his history of discontinuing treatment for schizophrenia and killing the wrong person. At an appeal, a second judge was persuaded to reduce the sentence by another year in light of new testimony from psychological experts about MAOA, which the presiding judge notes:

> In particular, carrying the low activity MAOA gene (MAOA-L) could make the subject more prone to express aggression if provoked or socially excluded. It should be stressed that such "genetic vulnerability"[1] turns out

> to carry even more significant weight in cases in which an individual grew up in a negative domestic social context, and was, especially in the early decades of life, exposed to adverse, psychologically traumatic environmental factors.[2] (*Bayout v. Francesco* 2009)

Arguments were brought forward that the accused's immigration from North Africa and social exclusion in his new environment in Italy should be counted as an adverse environment.

In *Waldroup* (2009), in the context of extreme marital strife, the defendant shot and killed a friend of his wife's and violently assaulted his wife, then took her hostage (she was rescued shortly thereafter by the arrival of police) (Hagerty 2010; Polk News 2009a, 2009b). Though the presiding judge claimed the evidence did not influence the decision to find Waldroup guilty of manslaughter (maximum sentence of 6 years) rather than murder (a capital offense), testimony regarding MAOA (and Waldroup's history of being abused as a child) was given during the guilt phase of the trial. Waldroup was also found guilty of aggravated kidnapping and attempted first-degree murder (of his wife), and received a total sentence of 32 years.

In *Albertani* (2011)[3], the accused killed her sister after having kept her sedated with benzodiazepines (tranquilizers) for several days; sometime after the disappearance of the sister, the accused was arrested when police arrived at her mother's house in response to screams overheard via their radio monitor to find the mother unconscious and on fire on the kitchen floor (*Albertani* Gip di Como, 20.08.2011; Feresin 2011). At the trial, there were conflicting and opposite psychiatric evaluations of the accused, prompting the judge to allow behavioral evidence and neuroimaging and genetic biomarker testing results to be brought forward; MAOA was one of the many genetic variants tested. She was found guilty of the murder of her sister and attempted murder of her parents, but her sentence was reduced by 10 years (from 30 to 20, 3 of which were designated to be at a psychiatric treatment facility) by the combined weight of the psychiatric, behavioral, and biomarker evaluation. As existing MAOA gene-environment interaction studies have mostly considered males, Albertani, being female, represents an example in which the application of the finding lies well outside the group in which it was demonstrated, highlighting the need for further research on diverse groups (and a point that is of current relevance to the "fit" of the evidence, which I address later in this chapter).

The invocation of the MAOA biomarker to reduce legal sentences has often been met with strong moral opposition, as it has been in these three cases. Though there are good reasons to temper the exculpatory power of this specific biomarker, the reasons often put forward are more general rejections of biomarkers of increased risk. I consider some of these general objections later in this chapter, where I examine whether probabilistic predispositions are legally relevant; I put forward a simple framework combining Bayesian statistics with the law's existing principles of folk psychology, which courts should be able to use to determine when a biomarker, set of biomarkers, or bio-actuarial prediction should rightly reduce legal responsibility (and when it should not). I argue that probabilistic predispositions (regardless of whether indicated by biomarkers or other sources) are relevant insofar as they change our degrees of belief about the events or mental states that are relevant to the legal questions at issue. By reconceptualizing the legal process as a whole as a process of Bayesian updating, in which the probabilities of certain propositions being true are increased or decreased depending on the evidence presented, I show that there is nothing categorically different between biopredicted risks and quotidian evidence; rather, they both can be weighted with the same logical methods (with the prior, i.e., default most likely proposition, that the events and mental states are such as would support innocence as dictated by the principles of criminal law).

PROPOSITION A: PROBABILISTIC PREDISPOSITIONS ARE LEGALLY RELEVANT

Are probabilistic predispositions, whether of the behavior of a person in a car or the behavior of a person in society, relevant to criminal trials? Though the quality of the answer to this question may change depending on the jurisdiction, it is not difficult to see that the legal structures are biased toward a similar sort of binary, categorical thinking as we saw with the concept of disorder. For example, any evidence submitted to court is first deemed admissible or not admissible, a binary framework that often leads courts to elide quantitative aspects. David L. Faigman comments on this dimension-reducing tendency in the context of the complexities inherent in determining the error rate of a technology:

> [. . .] the admissibility decision is concrete and categorical, while the complexities associated with the sources of possible error tend to cut in many directions. Courts have little time or inclination to explore all of the details of an admissibility decision and are prone to justify their categorical decision in categorical terms. (Faigman 2013, p. 106)

Many elements of other domains of the legal process, too, are framed in categorical terms. Finders of fact work to establish the existence or nonexistence of discrete facts, both empirical facts—were there or were there not skid marks?—and mental-state facts—was the defendant intoxicated or not intoxicated? Even when blood alcohol content (a biomarker with which many readers will be much more familiar than with the MAOA gene-environment interaction) is used in relation to the latter question, the output is often rendered categorical—over the legal limit or under—even when many would readily acknowledge that the relation of blood alcohol content (BAC) to the impaired cognitive and motor faculties that the law (and society) cares about is probabilistic; one can be impaired and under the limit or unimpaired and over the limit (especially when one considers variability in the timing and instrumentation for the measurement). BAC often is nonetheless useful in that it gives us probabilistic data, based on the rate of functional impairment in groups, of the likelihood of relevant impairment in the individual case.

One could interpret the US Federal Rules of Evidence's guidance on relevant evidence, for example, in a similar probabilistic sense. These rules describe evidence as relevant if it has "any tendency to make the existence of any fact that is of consequence to the determination of the action more probable or less probable than it would be without the evidence" (Federal Rules of Evidence, as quoted in Faigman 2013). Insofar as a predisposition (whether assayable from the model of car or from biomarkers) can meaningfully inform the probabilities of the truth of the beliefs held by the judge and jury, the predisposition is relevant.

The basic idea that beliefs come in *degrees*, and that the degree to which a hypothesis should be believed can be conceptualized as the probability one assigns to that the hypothesis being true, sits at the core of the field of Bayesian epistemology (mathematically formalized in Bayesian statistics). To illustrate degrees of belief, a juror may (1) believe that a defendant intended to kill a human being, and (2) believe that a defendant killed a human being, but assign a greater probability to the belief 2 than to belief 1. Degrees of belief are increased or decreased in an iterative process called Bayesian updating, which, stated in simplified form, is the process of modifying one's degree of belief in a hypothesis as a function of the prior probability of the hypothesis being true and the probability that the hypothesis is true given the new evidence (see Horwich 1993). This process seems to map reasonably well onto the US Federal Rules of Evidence statement of the relevance of evidence, and thus fits reasonably well in the legal setting.

The "Bayesian" conceptualization of probabilities is notably different from the probability theory that is taught most often, the "frequentist" one, in which probabilities are frequencies of particular events in aggregate groups. There is a long theoretical debate between Bayesian and Frequentist statisticians that is beyond the scope of this book, and there are strengths and weakness to each. Despite its limitations, Bayesianism can be very useful, in the words of Paul Horwich, "[. . .] if the point is to solve the many puzzles whose origin is the oversimplification that belief is an all-or-nothing matter [. . .]," which may be the case with biomarkers and the law. It also provides an intuitive way of thinking about how probabilities can apply to individuals.

Bayesian thinking, furthermore, might be slightly more straightforward inside the criminal courtroom than outside it. Outside the criminal courtroom, the priors (baseline beliefs) are often difficult to conceptualize as they are determined by all sorts of background and bias (which we might refer to tongue-in-cheek as "common sense"). Inside the criminal courtroom—in principle at least—the prior should be a high-probability assignment to a hypothesis of *innocence*. The overarching formulation in the courtroom might be restated as the probability that the defendant is innocent given the evidence.

If we shift from thinking in terms of establishing the presence or absence of discrete facts to modulating the probability that our beliefs (about events and mental states) are true, then we will rescue probability from the shadows of the legal process and potentially enable the relevance of predispositions to be weighted just like other evidence. In this context, it can be argued that predispositions are only as relevant as they shift the degrees of beliefs in the case.

OBJECTIONS: THE GROUP-TO-INDIVIDUAL PROBLEM AND THE PREDISPOSITION-TO-EVENT PROBLEM

There are likely to be several objections to my importation of Bayesian statistics to accommodate biomarkers in the courtroom: (1) the group-to-individual problem, and (2) the predisposition-to-event problem. In the group-to-individual problem, proponents argue that evidence gathered from groups (aggregated) cannot be applied to individuals. The predisposition-to-event problem is a variation on this, where proponents argue that predispositions cannot be applied to individual events, and so I will consider the two objections together. To put them more explicitly within legal terminology, both of these problems are examples where the "fit" of bioprediction in courts (i.e., whether it applies to legal issues) is called into question (Faigman 2013). I will first consider the group-to-individual problem.

The group-to-individual problem is sometimes stated with little more nuance than that with which I have introduced it here (as if its truth were indisputable), but it is also often put forward in more robust form, as illustrated in the following excerpt from Campbell and Eastman (2014) on the limits of studies looking at bioprediction of violence:

> However, even if a sufficiently large sample is used, the outcome [of many studies] is expressed in terms of averages based on group data. *This is in*

itself problematic as the law is concerned with evidence regarding a specific individual. And it is not the case that specific inferences, or predictions, regarding the individual can validly be drawn on the basis of observed statistically significant differences between a group to which the individual belongs and a relevant control group. Indeed, the differences with regard to the variable in question may be greater within groups than between experimental and control groups. (Campbell & Eastman 2014, p. 97, my emphasis).

My interpretation of Campbell and Eastman's argument is as follows:

a. Often, these studies have small sample sizes, and this limits the confidence one should put in the conclusions, but the group-to-individual problem *is independent of these sample size concerns.*
b. Valid inferences categorically cannot be made about individuals from this type of group data (statement of the group-to-individual problem).
c. This type of data is uninformative partly because the spread of the observed outcomes (e.g., incidence of violence) is very large within each group, and the difference between groups is small.

The group-to-individual problem in its categorical form makes large claims on the limits of the relevance of experiments on groups to the law. Of course, using group data to say something about individuals who are part of those groups is not a new business by any means. The practice of medicine, for example, is concerned with evidence about individual patients. Clinical decisions are many times made based upon inferences about individual patients based on their memberships in groups. The most well-known groups of this sort are the categories of disorder. Earlier in this book I argued that disorders are primarily heuristic categories that capture a risk of harm associated with biological variation. Applying information on groups to individuals, therefore, should be as morally problematic in medicine as it is in law. My hypothesis, however, is that few clinicians hold it to be categorically problematic, as asserted by the strong version of the group-to-individual problem.

The business of applying data on aggregates to individual cases is something that all of us do when we make decisions about single days and locations based on the weather report. To preempt the counterargument that decisions based on weather prediction are not as serious as those made in criminal courts, remember that the forecast is not just for barbecues and ball games. In 1944, for example, the meteorologist James Stagg advised that a beach event be postponed from June 5 to June 6 on account of higher uncertainty in the weather on June 5; of course, the event was D-Day, the Allied invasion of Normandy in World War II (as described in Ross 2014, and summarized in Rice 2014).

In thinking about whether group data can be relevant to individuals, Monahan (2014) finds compelling (as do I) the Russian roulette thought experiment put forward by Grove and Meehl (1996, 305–306). Like most thought experiments, this one involves a mad dictator forcing you to make a choice:

> Two revolvers are put on the table and you are informed that one of them has five live rounds with one empty chamber, the other has five empty chambers and one live cartridge, and you are required to play Russian roulette. If you live, you will go free. Which revolver would you choose? Unless you have a death wish, you would choose the one with the five empty chambers. Why? Because you would know that the odds are five to one that you will survive if you pick that revolver; whereas the odds are five to one you will be dead if you choose the other one. Would you seriously think, "well, it doesn't make any difference what the odds are. Inasmuch as I'm only going to do this once, there is no aggregate involved, so I might as well pick either one of these two revolvers; it doesn't matter which"? (Grove & Meehl 1996, 305–306; quoted in Monahan 2014, pp. 65–66)

Even if the case were modified to be a decision based more explicitly on group data, I think it would still have intuitional force:

> Imagine you are held hostage by a disgruntled statistician and forced to choose similarly between two revolvers. The statistician informs you that 50 people have played this game before you. Every game, he loads the revolvers the same way and places the revolvers in the same place. He informs you that, though the spread of outcomes within each group is wide (some die and some live in each group), on average those who chose the gun on the left were statistically more likely to die (and less likely to live) than those who chose the gun on the right ($p < 0.05$).

Even though the within-group difference is bigger than the between-group difference in probability of death, the gun on the right still seems like a reasonable choice. We can also connect this back to our Bayesian thinking. The prior belief that the revolvers are equivalent is made less probable by the evidence that they have behaved differently in groups; the degree of belief that choosing the right gun (the gun on the right) is safe is increased.

If the categorical version of the group-to-individual problem can be rejected, then the question becomes not whether group data can be applied to individuals that are part of groups, but the *degree* to which that application can usefully inform the probabilities of the events and mental states that the structures of legal responsibility care about. Reiterated by Morse and Newsome (2013), "a crucial issue that will frequently arise is the degree to which we can make valid inferences about an individual defendant from group data for a group of which the defendant is a member" (Morse & Newsome 2013). Within our Bayesian framework, "valid," I think, is best interpreted here as a probability of true belief surpassing a certain threshold. What that threshold is or should be may become a more pressing question as we become more sophisticated about how we quantify and apply these probabilities.

DEGREES OF APPLICATION: THE WEAKER VERSION OF THE GROUP-TO-INDIVIDUAL PROBLEM

With the strong version of the group-to-individual problem on the sidelines, there is still a challenge from a weaker (and more plausible) version: that the application to the individual from the group data wouldn't be very good at all in the context of law (even if it can sometimes be helpful in these other contexts). Consider, for example, this claim about a dichotomy between types of probabilities in science and in law:

> Policymakers may be interested in probabilities, but the court of law is not probabilistic in the same way that science is [. . .] the legal system considers the probability of a given individual being guilty as well as likelihood that he or she might reoffend. While they may use population or group data to inform this judgment, this is very different from behavioral genetic research. (Viding & McCrory 2014, p. 166)

While acknowledging a role for probabilistic thinking in the law (though not to the extent to which I am advocating), the implicit implication here seems to be that the type of group data from behavioral genetic research (e.g., the MAOA gene-environment interaction) is not compatible with the probabilities the law cares about. From where could this incompatibility stem?

One possibility that is seen frequently in the literature was raised by Campbell and Eastman (2014), in their articulation of the group-to-individual problem, which I interpreted as "C": This type of data is uninformative partly because the spread of the observed outcomes (e.g., incidence of violence) is very large within each group, and the difference between groups is small. If this "spread" is problematic and is observed often enough in these studies to consider it inseparable, then it could form the basis of an incompatibility argument. These two claims are indeed endorsed by other writers, for example, "it is a universal finding that the confidence interval around the mean score [a statistical measure of within-group spread] on any test or marker being investigated [as relevant to the law] is very wide. That necessarily means that the generalization from mean levels in two populations may not work as well as an individual predictor within any population" (Rutter 2014).

It does certainly seem that such wide spreads within groups could be problematic for claims on individuals; however, since the outcomes of the studies in question are often themselves binary (e.g., violent offense or no violent offense), we would expect the spread to be large. As pointed out by Monahan (2014), "If we use dichotomous variables the spread of outcomes will take up the entire range and the range of individuals will be by definition overlapping (Monahan 2014; also see Hanson & Howard 2010). Note as well that we already had the intuition from the revolver scenario that in cases of decision-making under uncertainty, information on probabilities with overlapping spreads could still be helpful (not as helpful, of course, as if the spreads were not overlapping, or overlapped less). So this objection, too, seems less forceful than is often put forward.

Others rephrase the "spread" objection in slightly different terms, that is, they observe that many or the majority of those in the biomarker group (e.g., MAOA combined with maltreatment) with a higher average rate of the outcome (e.g., violence) *will not go on to commit violence* and assert that this is morally problematic (for an example, see Forzano et al.'s [2010] commentary on the first Italian trial, *Bayout* [2009], to invoke the MAOA biomarker). Remember that we discussed this sort of objection before during the discussion of false positives and false negatives in the reorientation of disorder around risk in Part I of this book. There, as here, this argument has force primarily against the distillation of the probabilistic claims of bioprediction to absolute claims—that by membership in this "biomarker positive" group "it's just a matter of time" before the person commits violence (or develops profound cognitive impairment). If we keep the probabilistic claim in the foreground, on the other hand, we can weigh the evidence appropriately according to the statistical framework we deem most appropriate. The language of false positives and false negatives is commonly associated with Bayes's theorem (the same Bayes as the Bayesian degrees of belief), and though most often invoked in frequentist terms (the *number* of false positives and negatives in a population), it can usefully be thought of as evidence for how we should update the prior *degree of belief* that this person *x*; observed cases of false positives and negatives are useful insofar as they improve our predictive model of true belief (reflected by the probability we assign to the belief that *x*). Sometimes, this analysis will lead us to conclude that our predictive models do not improve if we apply a biomarker or set of biomarkers (e.g., the prediction from the raw prevalence of a legally relevant condition in a group is better than when the biomarker is applied to that group, as is sometimes the case when the condition is of very low prevalence) (Singh & Sinnott-Armstrong 2014).

Determining the probability that we should assign to individual beliefs that are important in the determination of legal responsibility, however, should often take more work than merely gleaning the numbers from a paper or meta-analysis. At the very least, in the law (as well as in the clinic) there is often uncertainty in the degree to which we should think of this particular individual as a member of the group in which the effect of the biomarker has been shown. This brings us back to the "application" component of "fit" and is likely to be a difficult business: "While courts may under-appreciate the complexities of the science in the applied science proffered in court, they largely ignore—and are sometimes downright hostile to considering—the 'applied' component of 'applied science'" (Faigman 2013, p. 106). Nonetheless, the question should be best interpreted in a probabilistic, not dichotomous, framework, as it will always be true that there is some probability that the person is not part of the group (even if he or she possesses a certain biomarker), for example, a boy with MAOA-L and abuse who wasn't raised in the same sort of socioeconomic or cultural environment as those on which the effect has been shown. Sometimes the shift in probabilities of belief that a biomarker might affect is so small that the decision to exclude it from consideration might be made, but still, this should be viewed as a decision based on probability.

If my arguments so far are convincing, then we have reached a place where probabilistic predispositions might be treated similar to other evidence when determining their relevance to legal questions. There is nothing categorically wrong about using group data speak to the individual (and sometimes it can indeed be very useful); rather, the question is the *degree* to which it can do so, an interrogation that might be usefully approached within a Bayesian framework.

It is certainly not true that scientific data on groups will always meaningfully change the probabilities relevant to the case. As Buckholtz and Meyer-Lindenberg nicely phrase it, "[g]roup- or population-level scientific data are relevant to the law only to the extent that the data bolster or weaken the evidence provided in an individual case" (Buckholtz & Meyer-Lindenberg 2014, pp. 143). The real question is *under what circumstances* can bioprediction be useful in determinations of legal responsibility, and *how much*, which brings us to examine proposition B.

PROPOSITION B: KNOWLEDGE OF RISK SHOULD CHANGE HOW OTHER EVIDENCE IS *READ*

In the case of the ignition switch with a small probability of shutting off the power to the car (and thus the power steering and the brakes), the knowledge of the predisposition would not make open-and-shut cases out of all car crashes involving that model, nor would it have the same evidentiary value in all cases. Neither should it be assumed that biological predispositions hold that potential for criminal cases in which they might be invoked. Indeed, it is very improbable that *any* single piece of evidence—of whatever sort—would have such overwhelming effect in a criminal case. Rather, the predisposition might enter as one voice in dialogue with the other circumstances, and the voice's volume and timbre would change according to the topic of conversation.

One potential role for bioprediction in the legal setting would be to change how other evidence is interpreted, which could sometimes reduce responsibility. This might happen in two related ways. The first would be that the predisposition could reduce the probability that would otherwise be assigned to a belief supporting culpability (increase the degree of doubt), such as degree of intoxication while driving. The second would be that the predisposition could increase the probability that would otherwise be assigned to a belief supporting an event or mental state relevant to a positive (partial or full) excuse or mitigation, such as relevantly altered cognitive state or provocation. It is my assessment that the first possibility is under-discussed in the existing literature, which focuses on excuse and mitigation.

To illustrate the first possibility, consider that in most car crash cases, details such as a lack of evasive action and skid marks, if they were submitted as evidence, would be interpreted to increase the probability of severe intoxication, as was initially done in Candice Anderson's case. But the probability a juror should assign to the belief that the defendant was intoxicated during the crash might be reduced if those same pieces of evidence supported an alternative—namely, that the car lost power, making steering and braking much more difficult. If structural brain imaging had revealed a type of blood-vessel malformation and studies had shown that groups who had this sort of biological variation have an increased risk of absence seizure (see Appendix II, Seizure Prediction), then again, the probability a juror should assign to the belief that the defendant was highly intoxicated during the crash might be reduced. How much the probability in that belief should be lowered can be meaningfully informed by the statistical data we have on the group of cars of which the defendant's is a member. How much more likely was losing power if one drove this make and model than another make and model? What about the blood-vessel malformation? How much other evidence is there that could make this possibility more likely?

To illustrate the second possibility, consider that classical diagnosis of a psychiatric condition is sometimes put forward in criminal cases as relevant to legal responsibility. Again, remember that inferring something about the individual from his or her membership in a diagnostic category is a form of bioprediction (Singh & Sinnott-Armstrong 2014). As I pointed out in Baum (2011), however, not all diagnoses are equally relevant to legal responsibility. First, different diagnostic categories signal different degrees of relevance according to their risk band for potentially legally relevant symptoms; a diagnosis of attention deficit disorder might have less impact on the probability of an exculpatory criterion than the presence of schizophrenia because each has a differential probability of a legally relevant symptom, for example delusions. This property is sometimes framed as the principle that the "severe" mental disorders are more likely to be relevant than the less severe mental disorders (Morse & Newsome 2013). Second, even when we acknowledge the heterogeneity of symptoms within severe disorders, symptoms like delusions are themselves only probabilistically related to legally relevant mental states; for example, delusions can sometimes signal that a defendant did not intend to kill a human being (e.g., he intended to kill an alien), or not (he believed he would be paid several thousands of dollars for killing a human being). This analysis is in line with the observation that most of the argument surrounding mental disorders is not the presence or absence, but "the content and severity of the signs and symptoms" (Morse & Newsome 2013). In all cases, however, independent circumstantial evidence or testimony would likely be presented to argue that the diagnosis/symptom/content is of the type that would increase the likelihood of a legally relevant mental state, for example, unusual behavior or speech near the time of the crime or a psychiatrist claiming the defendant had been in treatment for delusions of the type that would be legally relevant—thus increasing the likelihood that the defendant had a legally relevant delusion during the act.

Predictive biomarkers, just like any evidence, however, would be relevant to judgments of legal responsibility only insofar as they provide non-redundant information. Forzano and colleagues (2010, p. 519) bring up the valid point, that "genetic variants associated with schizophrenia do not add to the evaluation of the phenotype itself" as far as relevance to legal responsibility goes. A diagnosis of schizophrenia would rule out the usefulness of biomarkers associated with future diagnosis of schizophrenia, as the diagnosis itself signals a higher risk band in relation to the legally relevant disruptions of reasonable decision-making. If the biomarker gives new information on the likelihood of a legally relevant state, however, it would add value. For example, if a predictive biomarker was significantly associated with delusions particularly pertinent to responsibility (e.g., delusions of persecution or immediate threat as opposed to non-threatening delusions), this information would be legally probative above and beyond the diagnosis of schizophrenia. In this case, the biomarker would have a tighter probabilistic relationship to the element of the diagnosis that is relevant to legal responsibility. Biomarkers whose effects are relevant via a mechanism distinct from schizophrenia, which may be the case with the MAOA gene-environment interaction (as I will review later), would also be useful. In all cases, however, it should be questioned whether the addition of one or more biomarkers (or, indeed, any evidence!) to the model improves the model's prediction of the likelihood that the legally relevant deficit was indeed present; this is similar to current guidance on the use of predictive biomarkers in cardiovascular medicine, where it is suggested that incorporation of additional biomarkers into a predictive model is most useful when the biomarkers are uncorrelated with each other (Wang & Circulation 2011).

As I argued in Part I, one of the most important components of a disorder is that it gives us information about an increased likelihood of future harm to self. The various events, signs, and symptoms of a disorder were important in that context insofar as they were related to risk of harm to self. In the legal context, we can adapt this earlier conceptual work if we observe that the signs and symptoms of a disorder are important insofar as they are related to a particular condition or capacity that is relevant to legal responsibility. So the value of a diagnosis is determined by the magnitude of the increased likelihood of that legally relevant condition or capacity. Biomarkers, or sets of biomarkers, should be viewed the same way.

PROPOSITION C: EVIDENCE OF RISK SHOULD SOMETIMES LEAD TO A REDUCTION IN LEGAL RESPONSIBILITY

What types of risks, as suggested by biomarkers or sets of biomarkers, might be more likely to rightfully support a reduction in legal responsibility? In discussing the intersection of bioprediction and the law, many consider and object to the relevance of the direct bioprediction of antisocial or violent behavior (e.g., Campbell & Eastman 2006; Forzano et al. 2010). Some of these objections fall under the category of rejecting the legal relevance on the grounds *that it is probabilistic or applies group data to individuals*, styles of objection that my arguments earlier in this chapter have hopefully put to rest. Another style (and much more convincing) objection was alluded to in our discussion of proposition B, and is well voiced by this excerpt from Morse and Newsome (2013):

> [. . .] Causal knowledge, whether from neuroscience or any other science, can enhance the accuracy of behavioral predictions, but predictability is

> also not per se an excusing or mitigating condition, even if the predictability of the behavior is perfect [. . .] Even if the explanatory variables that enhance prediction are abnormal, *excuse or mitigation is warranted only if a genuine excusing or mitigating condition is present.* For example, recent research demonstrates that a history of childhood abuse coupled with a specific genetically produced variant in neurotransmitter metabolism [i.e., the MAOA gene-environment interaction] vastly increases the risk that a person will behave antisocially as an adolescent or young adult [. . .] no excuse or mitigation is applicable merely because these variables make antisocial behavior far more predictable. If the variables that enhance prediction also produce *a genuine excusing or mitigating condition, then excuse or mitigation is justified for the latter reason* and independently of the prediction. (Morse & Newsome 2013, pp. 155–156, my emphasis)

By "genuine excusing or mitigating condition," Morse and Newsome mean the mental states that exist in the excusing or mitigating legal structures, like automatism, insanity, provocation, and so on. Restated in probabilistic terms, the MAOA biomarker might increase the probability of antisocial behavior but say nothing about the probability of a legally excusing or mitigating mental state. Similarly, possessing a handgun might increase the likelihood of gun-related violence without saying anything about one's legal responsibility. I agree with the thrust of this argument, but I do not think that genuine excusing or mitigating conditions should necessarily be thought of as independent of the prediction of violence, especially in the case of the MAOA gene-environment interaction.

Sometimes the very same mechanism that increases the likelihood of violence is itself exculpatory—or partially so. This is the case with the partial excuse of provocation—being provoked increases the likelihood of violence while at the same time being considered a partially excusing condition; the prediction of violence is not entirely independent of the excusing condition. I previously have argued that in certain cases, the MAOA gene-environment interaction might rightly increase the likelihood that a partial defense of provocation would apply, *because the defendant may be more likely to be susceptible to provocation in a legally relevant way*, especially if certain gaps in the scientific literature are filled (Baum 2011).

Briefly, there are two main relevant bodies of research on MAOA. The first is the one on the MAOA gene-environment interaction that showed that the boys of the cohorts assayed who had MAOA-L and were maltreated had an increased risk of antisocial behavior, which I reviewed earlier in this chapter. The second is a set of imaging and behavioral studies that suggest that males with MAOA-L have more intense cognitive reactions to frightening or aversive stimuli and decreased control over acting on those reactions, a set of results that prompted Buckholtz and colleagues to propose at the end of an informative 2008 review of the work that "MAOA genotype modifies an individual's 'socioaffective scaffold': the basic neural equipment for social and emotional experience, which subserves cognitive routines for contextualizing social signals, decoding ambiguous social interactions and regulating affective response in the face of perceived interpersonal threat" (Buckholtz et al. 2008, p. 127). The difficulty is that these two bodies of work—the increased risk of antisocial behavior and the putative mechanism—have not been convincingly linked. The hypothesis is that this "socioaffective scaffold" might be thrown by the gene-environment interaction even more strongly in the direction suggested, but this remains untested. As it stands, MAOA genotype alone has little if any effect on impulsive violence (see Buckholtz & Meyer-Lindenberg 2014) and the effect of the gene-environment interaction on violence risk has an unknown mechanism. Nonetheless, if future research does support the hypothesis, the partial defense of provocation may be more likely to apply in the case of MAOA-L plus childhood abuse; to see why, we will need to review some particulars of this partially excusing condition.

In UK common law, an appeal to provocation, if successful, would reduce a verdict from murder to manslaughter (see Padfield 2008). In this jurisdiction, the partial defense of provocation consists of one subjective and two objective tests. The first (subjective) refers to the likelihood that the defendant was subjectively provoked; if the defendant was in a tower with a rifle and shot a passing shopper on the street, the likelihood of this first hypothesis being true would be very low. The second two (objective) appeal to the reasonable man: If the defendant was provoked, would a reasonable man have been provoked, and would a reasonable man have done as the defendant did? The details of the partial defense of provocation vary from jurisdiction to jurisdiction, but one element common to them (and perhaps a common place where neuroscience may become relevant) is this appeal to a "reasonable man" standard (Morse & Newsome 2013).

Though this reasonable man standard is referred to as "objective," it is almost always "subjective" in some way; the reasonable man is not always a reasonable 5'10" man in his early thirties, but is often imbued with at least the age and sex of the defendant. In the United Kingdom, the reasonable man standard largely follows the guidance set up in *Camplin* (1978): "the reasonable man is a person having the power of self-control to be expected of an ordinary person of the sex and age of the accused, but in other respects sharing such of the accused's characteristics as they think would affect the gravity of the provocation to him" (as quoted in Padfield 2008, p. 181). A biological variation that increased risk of violence via an increased likelihood of lost self-control would not modify the reasonable man, for example, but one that increased violence risk via an increased degree of *perceived provocation* might. If the mechanism of the MAOA gene-environment interaction were to be shown to be a similar but augmented version of the imaging genetics results, then an argument might be put forward that this biomarker would "affect the gravity of the provocation to him" as evidenced by more intense cognitive reaction to aversive stimuli. If the level of self-control remains constant but the provocation is graver, than it might validly increase the likelihood that a reasonable man might have done as the defendant did (thus satisfying all three limbs of provocation). This might effectively "reclassify" borderline cases.

The qualities that might rightly modify the reasonable man, including whether to allow characteristics that might affect self-control, are a matter of hot legal debate in many jurisdictions (Morse & Newsome 2013). In the United Kingdom, for example, a 1997 ruling (*Campbell* [1997]) enabled the consideration of self-control in a reasonable man, "in recognition of human frailty," in the decision's words; this element of the reasonable man remained subjective for nearly 8 years, until it was overruled in 2005 in *A-G for Jersey v. Holley* (2005) (Padfield 2008). If a jurisdiction were to make the reasonable man subjective in a similar way to *Campbell*, the control component of MAOA might become relevant, as might evidence of frontal lobe dysfunction (e.g., a tumor, as was the case in the 1997 UK trial of *Luc Thiet Thuan v. R*).

Whether the reasonable man *should* be subjectivized in this way leads us to question why we consider provocation a genuine (partial) excuse. Morse and Newsome (2013, p. 171) observe, "[. . .] there is controversy about the best explanation for this doctrine, but the dominant explanation is that rage diminishes the defendant's capacity for rationality or capacity for self-control, and the rage was not the defendant's fault because an ordinary person might have been similarly provoked under the circumstances." If the decision in *Camplin* (1978) that the level of self-control could be modified by the age and sex of the accused is itself reasonable, then what we may mean by an ordinary level of self-control is ordinary for a particular group (e.g., 18-year-old boys vs. 60-year-old women). What about groups based on biomarkers other than the sex chromosomes and the degree of gray hair?

Probability and the Reasonable Man

When viewed through a probabilistic lens, the reasonable man standard is perplexing in another way. Even if the self-control of the reasonable man remained constant, if we were to present him with increasingly provoking stimuli, might it be reasonable to hypothesize that there would be a level of provocation at which he would yield to his rage some (but not all) of the time? A man of reasonable self-control does not have perfect self-control, after all. But if we think a reasonable man would have yielded *some of the time* to the provocation, is that relevant to legal responsibility? Even if we rephrase it in Bayesian terms that would assign a probability of roughly 50 percent to the likelihood that it is true that a reasonable man would have done as the defendant did? And what if a data set emerged that showed that 5 percent of ordinary 28-year-old men (or even prescreened for above-average impulse control on a go/no-go task) would be provoked to punch a person who spit on them? Would it matter if it was 50 percent? 80 percent? What threshold of probability do we set to accept that it is *likely enough* that the reasonable man would have been provoked? These thought experiments are often left to be settled according to the "common sense" of jurors. An alternative is that we could adopt a graded standard; I introduced one system for matching the appropriateness of an attitude/response to the magnitude of risk of a future event in Chapter 4: careful risk banding. If we wish to avoid radically restructuring legal standards in this way, perhaps we can still appeal to flexibility in sentencing to enable a more finely grained treatment of desert.

Bioprediction and Sentencing

Because in many jurisdictions (especially so in the US) being found guilty of a certain crime can carry with it a wide range in length of sentence, it is in sentencing that there might be an opportunity to treat responsibility more as a continuum rather than as an all-or-nothing quality. Uncertainty in the reasonable man, in the degree to which the group data fit to the individual, in the degree to which the imaging genetics of MAOA fit to the gene-environment interaction, can be weighted and the sentence modulated accordingly. This is partially the reason that I have advocated in the past that the MAOA biomarker might be most relevant to sentencing (Baum 2011). Currently, the particular sentence within the range of allowed length is determined by intuitions about the deserts of the defendant. Slobogin (2014) has recently highlighted that some empirical work exists on the topic of the reliability of sentencing based on intuitions of desert and shows them to be highly variable. If the uncertainty in inferences is quantified or quantifiable, as uncertainty in aspects of biomarkers is, these inferences might be able to be rendered more uniform than if based on intuition alone.

Punishment is justified not only by retributivist (based on desert) reasons, however, but also for deterrence and for the protection of society from dangerous individuals. It is sometimes proposed, therefore, that bioprediction, especially of violence, can "cut both ways" and lead to longer sentences. Though the balance of these goals of punishment may shift according to the political climate, it is interesting to note that in some jurisdictions (like the US), sentencing based explicitly on risk of violence is currently prohibited (though used extensively in deciding parole); an exception is made, however, for past criminal behavior, as it is thought that this metric speaks both to level of desert and to risk (Monahan 2014). The arguments I illustrated in Chapter 6 might apply especially forcefully in these jurisdictions, as foreseeability of increased risk may speak both to desert and to risk. With increased *awareness* of this predisposition to impulsive reaction combined with access to genomic sequencing, one may soon be expected to take precautions to protect others from risk of harm; whether the precaution consists of eating oily fish to decrease impulsivity (Hibbeln et al. 2006), or something more demanding, remains open to debate. In considering this possibility, however, it is important to note that there is little empirical work looking at whether the biomarker increases recidivism, and it may turn out that it does not.

THE FUTURE OF BIOMARKERS IN LEGAL RESPONSIBILITY

In this chapter I have argued that evidence provided by bioprediction should be treated similarly to other evidence. I countered objections that biomarkers—whether because they are probabilistic, require that data garnered from groups be applied to individuals, or often have wide intragroup spread—are categorically different from the evidence that jurors are every day asked to weigh and thus have little place in determinations of legal responsibility. Finally, I highlighted criteria within the partial defense of provocation and sentencing by which bioprediction might rightly lead to a reduction in legal responsibility, a proposal I illustrated with the MAOA gene-environment interaction. In making this set of arguments, I introduced the Bayesian notion that beliefs—whether about facts in a case, mental states, ultimate verdicts, or proper length of sentence—come in *degrees* that might be usefully thought of as the probability (likelihood) that one should assign to the belief being true (or false). I argued that within this Bayesian framework, the legal process can be thought of as a set of nested hierarchical probabilities where both my concept of disorder as a probability dysfunction and biomarkers are easily integrated with traditional evidence and legal process. Thus, if we think it will be too burdensome to rid the legal system of its many categorical structures, then we can still reconceptualize many of these categorical decisions as probabilistic in a Bayesian sense.

Within this probabilistic framework, the degree to which biomarkers are relevant to legal responsibility is governed by the strength of their claims on the likelihood of one of the legally relevant excusing or mitigating conditions, or the degree of doubt they introduce in the interpretation of other evidence.

Consequently, it is less important whether a disorder is present/absent than the probability it signals of legally relevant impairments (its risk band of legally relevant impairment); by making its risk bands apparent, therefore, a probability dysfunction actually has the potential to be more clearly and fairly applied in legal settings than any categorical concept of disorder. This interpretation conforms to a generally compatibilist approach to legal responsibility (e.g., Vincent 2013). Also, within this framework, biomarkers, like most evidence, will seldom if ever act as magic bullets that pierce the heart of truth, but their relative impact on legal responsibility will need to be weighed in the context of all of the other evidence of the case in question; the degree to which a biomarker might alter legal responsibility is likely to be highly variable on a case-by-case basis.

There are a few other principles that may be drawn from the analysis I have presented. The first is that an actuarial tool for violence prediction will only be probative to legal responsibility if it is tightly curated to include risk factors that are associated with violence *via a mechanism that increases the likelihood of a legally exculpatory or mitigating factor.* Bio-actuarial tools of legally relevant mental states (e.g., types of delusions, cognitive impairment, pain, etc.) might also be promising but have the drawback of not being part of a nested prediction that increases the retrospective likelihood of the condition.

The second is that if some of the assumptions made in the applicability or fit of biomarkers to legal questions is still troubling, then it should be all the more troubling that the vast majority of evidence proffered in a legal setting has not been or cannot be systematically evaluated to estimate the uncertainty it harbors. For example, when was the last debate about the evidence for and against the error rate associated with the claim that someone with a motive is more likely to be the killer than someone without? Or to have killed intentionally versus unintentionally? Or the most recent study on the likelihood that a reasonable man would be provoked? If the point of evidence is to meaningfully allow us to modify our prior, then we have been relying almost exclusively on common sense and the persuasion of skilled law professionals to do this Bayesian updating.

Though common sense will most likely continue to play a large role in the Bayesian updating that is done in the law, the development of biomarkers raises the question of whether we should move from this largely unstructured approach to a more actuarial approach, where the weights of evidence or uncertainty in application are quantified where possible and presented to the court as guidance for decision-making. Such a shift will likely face challenges, both from the legal tradition of deciding weights of evidence through the skill of argument and from the observation that potentially relevant science is not done with an application to legal constructs in mind.

Morse, in his introduction to the MacArthur Law and Neuroscience Project's *Primer on Criminal Law and Neuroscience*, voices this latter concern: "very few [neuroscience] studies directly investigate questions of interest to the law, and even those that are indirectly relevant have not been undertaken for that purpose and are therefore of attenuated relevance" (Morse 2013, p. xviii). These studies are conducted to investigate basic questions about how the brain works, and, if they are intended to be applied to any discipline, that discipline is likely medicine. There is indeed a growing body of neuroscientists with medical training trying to enable that application, but the law has no such army of multilingual champions: "unlike other related areas, however, such as mental health and law, there is no established discipline or conceptual and empirical research tradition that examined the relationship between law and neuroscience that could help answer the relevance question. There are few experts with sufficient expertise in both fields to engage in sophisticated research" (Morse 2013, p. xviii). The first step to creating such a discipline or research tradition would be establishing the financial supports to enable the training of "legal scientists," and to support their research. In the United States, there exist significant governmental and university-level investments to support the training of medical scientists (MD-PhDs).[4] Such programs could serve as models to support and expand JD-PhD opportunities.

Even with educational programs, however, sophisticated neuroscience research designed directly to address legal questions will not be sustainable unless there are grant or other financial supports for them directly. While industry, via pharmaceutical and medical diagnostics companies, shoulders a large portion of the financial burden for getting good data to apply science to medicine, there are no such industrial partners in sight for the law. The shift in emphasis at the US National Institutes of Mental Health to funding research not into the disorders as currently stratified by the *DSM-5* but into groups with shared altered function in cognitive and rewards domains (Research Domain Criteria, or RDoC), however, provides a framework that more closely approximates the risk bands that might be of direct interest to the law. This conceptual shift on the part of funders might lead not only to the creation of more neuroscience of indirect legal relevance, but also to funding opportunities for "legal scientists."

I will conclude this chapter with the thought that it is possible to envision a much more active future role for legal scientists than outlined here. Where uncertainty in "fit" exists, it is generally assumed that there is nothing to be done; "questions put within legal process may be 'incapable' of adequately testing the evidence, because it is evidence susceptible only to scientific challenge and not legal challenge; that is, because it is not fully possible to ask questions from one paradigm inquiring into another" (Campbell & Eastman 2014, pp. 105–106). The court can merely opine on the degree of relevance of a biomarker to the particular individual in question, taking into consideration all the unanswered (but empirically answerable) questions about any given biomarker. It is not entirely implausible to imagine, if the quantitative expertise came to be part of law chambers, however, that judges could require new analyses of existing data sets to better elucidate the fit of a particular biomarker or other neuroscience to an individual case; such a request would be similar to the requests from scientific reviewers during the publication process, and the timeline of many criminal trials (months to years) is comparable to the time given in the publication process for researchers to generate such data.

NOTES

1. As I mentioned in Chapter 2, I choose to focus in this thesis primarily on non-genetic biomarkers. I include MAOA because it is a gene-environment interaction.
2. This quotation has been translated from Italian, the original language of the document.
3. Because of the complexity and potential interest of this case, I have included the detailed version excerpted from Baum and Savulescu (2014) in Appendix III.
4. See http://www.nigms.nih.gov/Training/InstPredoc/Pages/PredocOverview-MSTP.aspx.

PART III

Biopredicton and Society

In Part II, I discussed the impact of the development of bioprediction upon obligations of the individual toward others. I showed in Chapter 6 the ability of bioprediction to expand the domain of other-regarding obligations that we are able to meet (and thus, be blamed for failing to meet). In Chapter 7, I showed the ability of bioprediction, in special circumstances, to reveal previously unrecognized influences on the extent of those obligations.

Because the act of holding responsible is a social act, the discussion in Chapter 7 foreshadowed the impact of the development of bioprediction on obligations of society to the individual, for example, what is the strength of a social obligation to amend judgments of blame in light of the development of technologies to visualize this previously unknown influence on behavior? Part III will directly address the impact of bioprediction on the obligations of society to individuals. As the ability to develop and use biopredictive technologies becomes feasible, it will no longer be a neutral decision for our social institutions to operate according to the status quo.

In Chapter 8, I defend a social obligation to develop and provide access to biopredictive technologies through an appeal to prioritarian and egalitarian concern for the worse off in the allocation of social resources. Most of our social institutions act without information that could be gained through bioprediction. Part II addressed the importance of this information to criminal and civil courts, through the lens of intersection with moral responsibility. As highlighted in Chapter 7, there are many relevant, unanswered questions about the MAOA gene x environment interaction, as there are about the range of other biomarkers for behavior. There may be some who will argue that biomarkers, as a class, create no social obligations until those questions are answered. Such a response, I think, misses the point. The alternative I support is that the information that could be gained about individuals is often of *moral value* (as opposed to clinical value) and raises questions about the moral justification of our current practice such that there is a social obligation to develop, improve, and provide access to biopredictive technologies. In Part III, here, I examine the possibility of such an obligation with respect to priority setting within the realm of distributive justice. I highlight (1) theories of justice that argue that we get more moral value from benefiting the worse off, even if there is no change in the *size* of the benefit; (2) that biomarkers are capable of revealing a way of being worse off and thus their use is supported by (1); (3) that benefiting the worse off requires having accurate information on who is worse off; (4) that there is some obligation on the part of institutions that distribute to gather that information in order to morally distribute resources; (5) that the types of information that institutions currently gather and weigh to determine who is worse off are not sufficiently different from bioprediction to justify the exclusion of biomarkers in this determination; and (6) that there is therefore some obligation on the part of institutions to develop and/or use biomarkers.

8

Bioprediction and Priority

One of the driving forces behind the development of biomarkers has been the goal of increasing the *size of benefits* we are able to give to people, for example, by early treatment or prevention. This chapter considers the inverse: by giving us reason to change *who* we benefit, bioprediction might increase the *moral value* of the benefits we are able to give. Diverse theories of justice argue that it is important whom we benefit. Egalitarian and prioritarian theories endorse giving priority to those who are worse off, even if doing so is less efficient than benefiting the better off (e.g., Crisp 2003; Dworkin 1981a, 1981b; McKerlie 1997, 1989; Parfit 1997, 2012; Rabinowicz 2002). While individual theories differ on the weight of, and sometimes the reasons for, that priority, the consensus between these theories is the view that there is greater moral value in giving a benefit to someone who is worse off than giving a benefit of equivalent size to someone who is better off. Such a basis for priority partly undergirds gathering money from the healthy to benefit the sick and providing social services to those disadvantaged in our societies. I appeal to this consensus priority view in hopes of showing the relevance of bioprediction across the entire range.

If one endorses this sort of egalitarian or prioritarian theory of distribution, it is crucial to have accurate mechanisms by which to identify the worse off; it would be at best a callous action and at worse a miscarriage of justice to distribute benefits preferentially to someone who was actually better off, all else being equal. What has traditionally been taken into account in the judgment of being worse off are many disadvantages, including poverty, disability, disease, and discrimination. The development of biomarkers, however, raises the worrying possibility that we have been operating in ignorance of an important way of being worse off, namely, being in a nontraditional way at high risk of harm, and have thereby been unwittingly distributing unjustly.

In this chapter I support the validity of this worrying possibility by defending against a series of objections the relevance of biological risk of harm to priority-setting. I show that bioprediction of future harm can show one to be worse off, not *merely at risk* of being so, even if the person in the high-risk category does not know his risk status and even if there is no clinical intervention able to diminish that risk.

I argue, moreover, that the bioprediction of risk of future harm (widely defined) should not be viewed as merely a "tie-breaker" in priority setting, but rather is at the core of our moral concern in a similar way that it is at the core of the concept of disorder. Although harm in the present matters morally, risk of future harm can in certain circumstances outweigh the claims of current harm for priority. I specifically exploit the conceptual overlap between bioprediction and disorders traditionally recognized to grant claims to priority (by making one worse off). I show that being identified as at risk of harm through traditional means (diagnostic pathways or socioeconomic demographics) or nontraditional means (biomarkers) is unimportant; rather, the magnitude and type of risk is what matters morally. I supplement this application of the Probability Dysfunction theory with a discussion of the moral relevance of risk concentration and physical security.

While our previous ignorance of predictive biomarkers may excuse the historical exclusion from considerations of priority of this way of being worse off, the excuse is no longer valid. Systems subscribing to priority, therefore, have at minimum reason to develop and provide access to biopredictive testing; if bioprediction is provided, not just encouraged, it will prevent the unjust and counterintuitive outcome of only the materially better off being able to demonstrate that they are worse off in health.

First, it is necessary to outline in more depth the theories that grant priority to the worse off.

EGALITARIANISM, PRIORITARIANISM, AND A CONSENSUS CONCERN FOR THE WORSE OFF

Each week in Boston, a converted Winnebago RV rolls into a handful of neighborhoods to provide free preventive health care and health education. A Harvard Medical School–affiliated program funded by donations and governmental grants, the "Family Van" aims to "remov[e] barriers to care and improv[e] the health of Boston's most vulnerable residents."[1] Although the Family Van has recently been shown to be surprisingly cost-effective (Oriol et al. 2009), its value is not usually understood in terms of its ability to reduce the cost of the healthcare system. The driving force behind the Family Van is a commitment to distributive justice: that we should work to eliminate the health inequalities between these underserved neighborhoods and other communities, even if we could attain similar cost-effectiveness by providing services to the other communities.

If we could have a similar benefit by providing services to other communities, what reason could we have for preferring to serve *these particular* communities?

We could reframe as follows:

> There are two equal-sized communities, A and B, where those in A are in general worse off than those in B.
>
> The Family Van can provide similar health benefits for the same operating cost to A and B.
>
> The Family Van can serve either A or B but not both.

Utilitarian theories would claim that we have no reason to prefer serving A over B; the cumulative benefits are the same in each case, and it is morally irrelevant who, in particular, is benefited. My purpose in this chapter, however, is not to argue against utilitarianism, but to show the moral relevance of bioprediction to those groups who do think that we have reason to serve A over B.

Such groups might be separated broadly into the egalitarian, the prioritarian, and the sufficientarian. My review of these theories will largely follow the structure and terminology laid out by Roger Crisp (2003), which I find to be exceptionally clear and intuitive; I will illustrate the application of that structure through variations on the Family Van. Consider, first, the egalitarian.

> The Egalitarian[2]: inequality (e.g., of resources, opportunities, welfare, capabilities) is itself bad. One state of affairs is to be preferred over another if there is less inequality. We should send the Family Van to community A because this will reduce the inequality between A and B. Sending the Family Van to B would increase this inequality and thus be an unjust action.

Now, imagine that I pointed out to the simple egalitarian that we could reduce the inequality between A and B to a slightly greater extent if we had the Family Van spread germs and health misinformation in community B (and thereby make B as badly off as A). The simple egalitarian would conclude that this course of action is better than benefiting A, because it obtains a state of affairs with slightly less inequality.

Nancy Oriol, the founder of the Family Van (and Dean of Students at Harvard Medical School), is unlikely to agree with the simple egalitarian. To what alternative view could she appeal in order to justify a preference for serving community A, even though this would lead to *more inequality* than misleading community B?[3]

To avoid this "leveling-down objection" of egalitarianism, Nancy Oriol could adopt the absolute priority view:

> *The Absolute Priority View*: when benefiting others, the worst-off individual (or individuals) is (or are) to be given absolute priority over the better off. (Crisp 2003, p. 752)

We are not so much concerned with equality itself as we are with the plight of the worse off. Equality in itself has no value, so we have no reason for misleading B. We should send the Van to A because they are the worst off. The absolute priority view would have us send the Family Van to community A even if doing so would provide very small benefit. Those who find this implication of the absolute priority view unpalatable might adopt the weighted priority view:

> *The Weighted Priority View*: benefiting people matters more the worse off those people are, the more of those people there are, and the greater the benefits in question. (Crisp 2003, pp. 752)

The weighted priority view would posit that there could be some level of greater health benefit to those in B, either more benefit per individual or lesser benefits to more people, which would justify sending the Family Van to B even though A is worse off. Such a view might underlie a belief that it would be wrong to send the Family Van to an even worse off community hundreds of miles away in Baltimore because the time and fuel spent in transit would mean benefiting fewer people.

The weighted priority view, however, would endorse giving priority to benefiting very large numbers of already very well off people by a very small amount over benefiting small populations of badly off people who could be benefited a lot. To avoid this outcome, one could add a sufficiency view. One example of a sufficientarian view is proposed by Crisp (2003):

> *The Compassion Principle*: absolute priority is to be given to benefits to those below the threshold at which compassion enters. Below the threshold, benefiting people matters more the worse off those people are, the more of those people there are, and the greater the size of the benefit in question. Above the threshold, or in cases concerning only trivial benefits below the threshold, no priority is to be given. (Crisp 2003, pp. 758)

As long as community A is beneath the threshold, sufficientarianism would still endorse sending the Family Van to A.[4]

All of these theories have at their core some version of what Parfit calls the "priority view."

> *The Priority View*: benefiting people matters more the worse off these people are. (Parfit 1997, pp. 213)

The moral value of the Family Van is inversely proportional to the well-being of those it benefits. Since community A is worse off, benefiting them is morally preferable. Even if we could have a greater health benefit to community B, the simple prioritarian could still argue that we should send the Family Van to A because the moral value of the greater health benefit would be outweighed by the moral value of helping the worse off.

In order for the arguments presented in this chapter to achieve maximum practical relevance in a morally pluralistic environment, it is the consensus concern for the worse off to which I will appeal to ground the moral value of bioprediction; I acknowledge that the arguments will have less moral weight on some theories, but even in those, it should have *some weight*. Despite the significant points of departure between these three major groups, the diversity within each, and their consequentialist and non-consequentialist origins, all of them agree that we should give priority to the worse off.

I will follow convention to refer to this consensus as the priority view (PV).

The egalitarian endorses the PV insofar as it reduces inequality (although inequality could be reduced in other ways), the prioritarians because a benefit has greater moral worth the worse off a person (though the priority may to some be absolute and to others defeasible by size of benefit to others or number benefited), and the sufficientarian insofar as those benefited are below some threshold level.

In cases where the size of the benefit and the number benefited are the same (and the worse off under the threshold), all these theories would agree that it would be wrong to preferentially benefit someone or a group of people that is actually better off.[5] To give priority to the better off over the worse off would be at best a callous action and at worst a miscarriage of justice.

Assuming the PV takes into account ways of being worse off that the distributor (and even the person in question) is ignorant of,[6] distributing resources according to the PV would encounter a practical epistemic constraint:

> *Epistemic Constraint*: in order to benefit the worse off, we need know who is worse off.[7]

The emerging possibilities of bioprediction show that we have been ignorant of a significant group of people who are worse off because they are in a nontraditional way at risk of harm. Bioprediction should be valued by proponents of the PV because it helps us to satisfy the epistemic constraint (to know who is worse off), and is thereby morally valuable *independent of the potential for bioprediction to produce health benefits.*

In what follows, I will defend the claim that those at risk of *future harm* are indeed worse off in a way that matters to proponents of the PV for priority in distribution *now.*

THOSE BIOLOGICALLY AT HIGH RISK OF HARM ARE WORSE OFF, WHETHER IDENTIFIED THROUGH TRADITIONAL DIAGNOSIS OR BY BIOMARKERS

It would be a relatively uncontroversial point to most theorists that those with disorder (the broad term, as introduced in Part I) are worse off in a way that matters to the PV.[8] One line of argument, therefore, is that bioprediction is relevant to the PV insofar as it is able to identify people that have some type of disorder.

I argued in Part I that disorder should be thought of as a probability dysfunction with the "risk clause" at its core; I questioned the defensibility of maintaining a distinction between classical disorders defined by clinical signs and symptoms and high risk states defined by predictive biomarkers. In Part II, Chapter 6, I illustrated this phenomenon when I argued that this concept of disorder disintegrates the distinction between a high risk of seizure soon that is identified by predictive biomarkers or by a classical diagnosis of epilepsy. I also argued that we should move away from a binary conception of disorder, that is, we should care less about whether someone has a disorder and more about the precise magnitude and type of future risk of harm; it is the shape of the probability function, stratified into risk bands, that should guide response. By doing so, we can sidestep the conceptual problem of distinguishing between very high predispositions to classical disorders and the classical disorders themselves; each should be thought of as a separate risk band (i.e., different disorders) with a different probability function of the same set of harms.

The application of this theory of disorder to priority is to posit that individuals in higher risk bands of harm are increasingly worse off and thus have stronger claims to priority according to the PV. We can see this illustrated intuitively when we judge that of two people with cancer, the worse off is the one whose cancer is more likely to lead to pain and death in the future. Of two people in pain, the worse off is the one more likely to experience pain in the future, even if the other has slightly greater pain now. Risk of future harm is equally relevant according to the PV, whether estimated by a diagnosis or by a set of predictive biomarkers.

More work needs to be done, however, before we can conclude that those at high risk of harm as identified by bioprediction are worse off in the relevant way for priority. There is no guarantee that my risk-reoriented definition of disorder shares all the properties of previous definitions, and as such, it is important to validate through independent means that a certain set of biomarkers can indeed show someone to be worse off in a morally relevant way. One might object that even if we accept these high-risk states as disorders according to my Probability Dysfunction model, this type of disorder does not actually make one worse off in a morally relevant way: it only increases the likelihood of one being worse off in the future (when and if one gets the "traditional disorder").

In another context, Norman Daniels has argued for the moral relevance of "Risk Concentration" in decisions to give distributive priority to someone with a higher risk of harm over many with lower risks of harm, even when the expected utility is the same between the parties (Daniels 2012a, 2012b). He argues that this risk concentration justifies giving priority to treatment over prevention. I examine next whether "Risk Concentration" can help clarify the relevance of bioprediction to priority.

Historically, concentration of risk has been shown by the manifestation of visible signs, symptoms, and sickness behavior. As bioprediction progresses, it will become increasingly possible to show concentration of risk by nontraditional means. If risk concentration is what undergirds priority, then bioprediction may collapse the treatment/prevention distinction.

Risk Concentration as Grounds for Priority

To examine whether risk concentration matters morally, Daniels proposes that we construct two "fully equalized cases" of the sort initially developed by Kamm (2007, 347–9) to examine whether distance matters morally in distribution. The goal of the technique is to construct cases where only the concentration of risk varies between them. Daniels proposes the following:

> Suppose we have only five tablets of a medicine that can be used either as an effective treatment for a [definitely terminal disorder], provided that all five tablets are given, or that can be given in one tablet doses to people exposed to the [disorder], where it acts as an effective vaccination. Without vaccination, there is a 20%chance of contracting the [disorder] once exposed.
>
> *Treatment*: Cory has the [disorder]. We can give her the whole dose.
>
> *Vaccination*: Dolly, Ellie, Fannie, Gert, and Hannah have been exposed to Cory. We can vaccinate each one of them with one-fifth of the whole dose we can give to Cory. (Daniels 2012a, p. 188; also see Daniels 2012b, p. 41).

Daniels explains that even though there will be one expected life saved in either case, he has a clear intuition that we should give Cory priority.

> My intuition depends on Cory having the stronger claim on assistance since she is, *ex ante*, clearly worse off than the other five. She faces certain death if nothing is done or if the others get a vaccination, whereas they face "only" a 20% chance of death if nothing is done or if Cory is rescued. If her claim on assistance is stronger, it would be unfair to her to favor the others, who have weaker claims. (Daniels 2012a, p. 189)

Daniels bases his preference of treatment over prevention upon a belief that we should give priority to the worse off. He has the intuition that Cory is worse off because she has a higher risk of death than the others. The claim for assistance is not due to having "the disease," therefore, so much as to the risk of death signified by the diagnosis. Since he grounds this judgment on the risk of death sometime in the future, Daniels would seem to be required to accept that a biological condition that puts you at high risk of death, even if it had no other effect, would make someone worse off in a similarly relevant way, even if the person had no traditional disease. He would probably maintain his position that Cory is worse off even if the other five were currently suffering moderate pain due to gout, which would suggest that the risk of harm in the future can outweigh present harm.

If we want to use this strategy of fully equalized cases to clearly get at whether risk concentration matters morally in bioprediction, we need to amend the treat/vaccinate case in two ways. First, we need to better control for any remaining intuitions that vaccination might result in fewer or greater than one death. Elsewhere, Daniels describes that many do not share his intuition to treat Cory (Daniels 2012b, pp. 41);[9] rather than saying anything about the moral value of risk concentration, this disagreement may exist because some people are at some level unconvinced that one person will die regardless of the choice we make. If we assume that each of the five individuals represents an independent "throw of the dice," for example, each throw having 20 percent chance of coming up dead, the binomial theorem estimates the likelihood of each outcome as in Table 8.1.

Table 8.1 CHANCES OF OUTCOMES IN "VACCINATION" ACCORDING TO THE BINOMIAL THEORUM

Outcome	0 Die	1 Dies	2 Die	3 Die	4 Die	5 Die
Probability	32.77%	41.96%	20.48%	5.12%	0.64%	0.03%

The intuitions of people who interpret the case this way may have nothing to do with risk concentration, and are complicated by difficulty in how to weigh multiple lives. Because there is a 25 percent chance that more than one person will die if we treat rather than vaccinate, some may think we should vaccinate to prevent the worst outcome. Others may treat Cory to have the best chance of the best outcome; if we treat Cory, she lives and there is a 33 percent chance that no one else will die.[10]

Second, we will want to better control for preferences for certainty over uncertainty. In most practical situations, we rarely deal with certainties, and indeed in the realm of biomarkers it is precisely the uncertainties that are interesting. Some might choose to give Cory priority simply because benefiting her is a certainty, while the benefit to the others is uncertain (after all, there is a chance that none will die if left unvaccinated).

To be maximally helpful in deciding the value of risk concentration in the discussion of bioprediction and priority, therefore, we need a case where there is (a) the same (uncertain) risk of one person dying regardless of whether we give priority to the one with more concentrated risk or not, and (b) *no possibility that more than one person will die*. Here is my attempt to make such a fully equalized case:

> Sally has a transmissible disorder, which is characterized by a set of innocuous clinical signs (e.g., smell of urine, enlargement of liver, subtle skin pigmentation), but carries a 90 percent chance of death (and 10 percent of uneventful resolution) if left untreated. Hans, Franz, Hansel, Gretel, Thor, Helga, Olga, and Billy-Bob were exposed to Sally and are at elevated risk of developing the disorder such that when we calculate the proper statistics, they each have a 25 percent chance of death unless preventively treated. There is thus a $(0.75)^8$ or 10 percent chance that no one will die if we treat Sally, just as there is a 10 percent chance that no one will die if we preventively treat the eight.
>
> We have eight aliquots of a medicine that is made from modified human antibodies. If all eight aliquots are given, they will be an effective treatment for Sally; alternatively, one aliquot each can be given to Hans, Franz, Hansel, Gretel, Thor, Helga, Olga, and Billy-Bob as effective prophylaxis. However, if we treat Sally, we need there to be no risk that *more than one* of the other eight will die; fortunately, should we choose to treat Sally and one person happens to die, we can quickly make enough medicine to treat the seven remaining people (56 aliquots) by harvesting, purifying, and chemically modifying (to increase potency) large quantities of human antibody from the newly deceased.

If Sally is treated, there is a 90 percent chance of one life being lost; if Hans, Franz, Hansel, Gretel, Thor, Helga, Olga, and Billy-Bob are preventively treated, there is a 90 percent chance of one life being lost. The possibility of quickly harvesting more medicine after a death should one occur frees us of the limitations of Daniels's earlier cases (i.e., we no longer have to weigh multiple lives, worst-case or best-case scenarios). There is no longer a risk of more than one person dying. We also no longer are comparing certain to uncertain benefits. Both benefits are uncertain to the same degree.

With these fully equalized cases, we are prepared to ask whether the concentration of risk matters morally. My intuition (and I would expect Daniels would agree) is that we have a greater obligation to treat Sally than to vaccinate Hans, Franz, Hansel, Gretel, Thor, Helga, Olga, and Billy-Bob, even though both courses of action have the same expected number of lives saved. Similarly, I base my judgment on the intuition that Sally is worse off than each of the eight who have been exposed.

What does the moral work is not that Sally is identified as having a disorder, and it is not that treatment should take priority over prevention. Rather, it is that Sally is at a higher risk of death than each of the others when evaluated individually. To see that risk concentration and not diagnosis or treatment is at work, we can reverse the scenario.

> Hans, Franz, Hansel, Gretel, Thor, Helga, Olga, and Billy-Bob have a disorder with a risk of death of 25 percent each and it is Sally who has been exposed to the eight. Sally, however, has a set of predictive inflammatory biomarkers associated with high rates of disorder conversion followed by death such that Sally's risk of death is 90 percent, *even though she does not have the disorder now*. We need all eight aliquots of the medicine to be confident of preventing the disorder in Sally, but Hans, Franz, Hansel, Gretel, Thor, Helga, Olga, and Billy-Bob only need one aliquot each for effective treatment. Again, there is no risk of *more than one person* dying if we choose to prevent the disorder in Sally because we can harvest more medicine, as in the previous case.

In this case, Sally is at higher concentrated risk (90 percent risk of death), but does not currently have the disorder and thus requires prevention. The others are each at lower concentrated risk (25 percent chance of death each) but have the disorder and require treatment. We should give priority to Sally because she has a greater concentration of risk of death (higher risk band), and is thereby worse off, even though she currently does not have the classical disorder and even though she requires prevention, not treatment. Moreover, we see that the intuition of Sally being worse off follows the concentration of risk, regardless of whether the risk is identified via biomarkers or classical signs. While it may be true historically that higher concentrated risk is found with classical symptoms and treatment, especially in mental health, the development of bioprediction will increasingly make this untrue. Indeed, blood glucose levels are often more informative of risk of harm than classical signs or symptoms of diabetes.

The PV, therefore, would have us give priority to Sally when she has the highest concentration of risk, regardless of whether she has a disorder or predisposition or requires treatment or prevention. An institution that felt it to be morally important to distribute according to the PV, therefore, should not be satisfied with a policy that gives priority to individuals whose risk concentration happens to be identifiable by classical diagnosis but not to those whose risk is identifiable only with bioprediction. Such statistical discrimination could make those whose risk is identified by bioprediction even worse off, and undermine the ability of the institution to meet the PV even if it were to act in accordance with the PV in the classical cases.

Risk of Harm and Being Worse Off

With the example of risk concentration, I built an intuitional case that being at high risk of future harm makes one worse off than being at lower risk, or than being at lower risk with some amount of current harm, in a way that should be given priority by the PV. Some who endorse the PV may not share this intuition. These opponents might exhibit the same style of binary thinking about disorder that we encountered in Chapter 5. Remember Sperling et al.'s (2011) statement about biomarkers of risk of dementia:

> Use of these biomarkers in the clinical setting is currently unwarranted because many individuals [. . .] may not develop the clinical features of [Alzheimer's disease] in their lifetime. Inappropriate use of this information in this context could be associated with unwarranted concern [. . .]. (Sperling et al. 2011)

What form might this binary thinking take in the context of priority? One possibility is the binary objection.

> *The Binary Objection*: the use of biomarkers of risk of harm in priority-setting is currently unwarranted because many individuals in Sally's position may not experience the predicted harm in their lifetime and thus may

> not be worse off. Inappropriate use of this information in the distribution of benefits may lead to the assignment of unwarranted priority.

When we first encountered this sort of binary thinking in Chapter 5, the proponents were concerned that, because the category is probabilistic, some in the category will, in truth, never go on to develop dementia, and thus there would be unnecessary treatment, stigmatization, and worry. I argued that we should reject this type of thinking because it sets an impossibly high epistemic standard that is completely inconsistent with our current practices; all disorders have probabilistic components of prognosis that we do consider.

To see that my argument generalizes to this context, consider this extreme rejoinder in the style of the binary objection: even if we know with absolute certainty (somehow) that Sally is, in truth, about to die, use of this information is currently unwarranted in priority-setting because many people who are about to die may otherwise have had years of life with such a low level of well-being that it would not be worth living. Thus Sally may not be worse off by being about to die.

This conclusion is implausible, and would not be defended by anyone affirming the PV. We should reject the binary objection as irrelevant to an action-guiding moral theory as well as to medical practice.

A more nuanced way to handle this sort of uncertainty is to think in terms of *prospects* and *expected value*, which tracks closely with the construct of *risk banding*. Consider the following "wheel of fortune" thought experiment adapted from Savulescu (2001, see pp. 414 for the original thought experiment):

> *Wheel of Fortune*: You are on the *Wheel of Fortune* TV show. You are given a chance to spin the wheel. The wheel is spun twice and the first value is placed into box A and the second value into box B, but you do not know what value is in which box. You are told that if you open Box A, you get the value inside, but there is a 50% chance that you will lose $100 from that value. If you open Box B, you get the value inside, but there is a 5% chance that you will lose $100 from that value.

If the game-show host tells you that you will open Box A and your opponent will open Box B, we might have the intuition that you are worse off for having to open Box A, *even though you might win more money.* This is because your prospects are worse. We can conceptualize how your prospects are worse by adapting a version of expected-value theory (Jackson 1991; Rabinowicz 2002); we multiply the probability of each uncertain prospect by its value. Since you and your opponent have no way of knowing what value is in each box, you have the same expected value other than the money you stand to lose. You have a 50 percent chance of losing \$100 and a 50 percent of losing \$0, so your expected value is 0.5 x (–\$100) + 0.5 x \$0 = –\$50; your opponent has a 5 percent of losing \$100 and a 95 percent chance of losing \$0, so her expected value is 0.05 x (–\$100) + .95 x £0 = –\$5. You expect to lose 10 times as much as your opponent, so your prospects are worse and you are worse off.

Another example would be a slanted playing field. If you are on a rugby team and playing a match on a pitch that is slanted against you (so you are looking uphill), then your prospects of winning the game and of getting through the game without an injury are worse. Your prospects of winning each scrum, ruck, maul, try, and line-out, are worse; and each time you lose one of these plays, your prospects of winning are adversely affected. You may still win the game, but you are (*prima facie*) worse off than if you had a level playing field.[11] Poverty makes one worse off, in part, because it lowers one's prospects. While it is certainly true that poor prospects and opportunities do not always result in lower welfare, we say that such exceptions "triumphed against the odds" or "got lucky;" they became better off than we expected them to be.

Biopredicton of risk of harm, similarly, gives meaningful information about prospects for health and is thus similarly important to the PV. It is not necessary for the PV to require certainty. Again, matching priority to risk band captures these ideas.

Knowledge About One's Own Risk of Harm

Couched within Sperling et al.'s (2011) statement, however, is a subtler challenge to the relevance of bioprediction to the PV. The authors are concerned that the *knowledge of risk* will cause worry. We might reformulate the essence of this concern into the *epistemic objection*.

> *The Epistemic Objection*: bioprediction of risk of future harm matters to the PV only if a person has *knowledge of his high risk of harm*.

The epistemic objection is challenging because its corollary would suggest that bioprediction, by giving individuals knowledge of risk, actually *makes them worse off*. Someone who embraces the objection, therefore, might argue that bioprediction should be neither developed nor provided.

To answer this challenge, I will look to the theory of well-being, and make the reasonable assumption that one is worse off if their well-being is less. According to *hedonistic theories*, how someone's life goes (his or her well-being) is determined by the overall balance of pleasure and pain he or she experiences. In *preference hedonism*, a pain would be an unwanted experience and pleasure an experience that is preferred; the amount of pleasure/pain is proportional to the degree it is preferred/unwanted. In *desire theories*, well-being is defined by the degree to which one's desires are fulfilled. In *objective list theories*, well-being is determined by the balance of objective goods and objective bads; examples of objective good would be having meaningful relationships, knowledge, development of one's potential; examples of objective bads would be being deceived, manipulated, and made un-free.[12] Pain would make one's life worse according to objective lists because it is an objective bad, according to desire theory if one has a desire to be free from pain, and according to preference hedonism if the pain is unwanted.

A preference hedonist may argue that Sally can be objectively at high risk of harm without having her well-being be less than someone at low risk. Since "what can be good or bad for someone can only be discernible features of [her] conscious life" (Parfit 1984, 500), according to this theory, Sally is not worse off if *she does not know that her prospects are worse.* If she knows that she is at high risk, this would provoke anxiety or fear about the future. If we had a case in which Helga, at a 25 percent risk of death, *knew* her own prospects, but Sally did not know she had a 90 percent risk, therefore, the preference-hedonist prioritarian would argue that we should actually give priority to vaccinating Helga. The preference hedonist gives absolute priority to self-knowledge of risk.

There is something unsettling about the preference hedonist's conclusion. Desire and objective list theory help us avoid it. Derek Parfit's *global success theory* of desire fulfillment permits desires about our objective prospects: "On this theory, what would make my life go best depends on what I would prefer, now and in the various alternatives, if I knew all of the relevant facts about the alternatives . . . [S]omething is bad for someone, only when, if he knew the facts, he would want to avoid it" (Parfit 1984, 499). According to success theory, well-being extends beyond what one is immediately conscious of. If, knowing the relevant facts about prospects, Sally would prefer to avoid being at high risk of harm, then it can affect her well-being without her knowing it. Because I can have desires about whether I am achieving what I have set out to achieve, and these temporally extended desires are global, then my objective prospects can affect the fulfillment of these desires. If, as I implied earlier, many people have a desire not to risk harming those they hold dear, my objective risk of having a seizure (or heart attack or other event) while driving my children to school affects whether I am achieving my desire. The objective list theorist would go further to argue that even if I do not have this type of desire, I *should.* Sally, even if she does not know about her prospects, therefore, would still be worse off.

There is one more argument against the epistemic objection, which stems from the literature on physical security and the value of true belief. If I believe that I am not at risk of death soon, but I am not right in this belief, then I cannot act as a "reliable planning agent" (John 2011). Since I have an interest in my ability to consider how to pursue my global desires/maximize objective goods (my ability to plan), and such an interest is dependent on this ability being reliable, physical insecurity (risk of harm) damages my interests by undermining this reliability. Well-being, here, is thought to be determined in part by *agency* rather than by *achievement*. Note that brain-based risks, which are risks to my reasoning capacity, would be especially able to undermine reliable planning agency, according this theory. Bioprediction of brain-based disorder, therefore, would be especially pertinent.

If we agree with Parfit, however, that each theory of well-being captures only part of what we consider well-being, then we should adopt a composite theory. I have shown here that being at high risk of harm, even if one does not know one's poor prospects, can make one worse off, unless one is *exclusively a preference hedonist*. In a composite theory, therefore, we should consider someone with poor prospects primarily to be worse off regardless of self-knowledge, but that the level of overall well-being can be tempered by the current cognitive state.

NESTED-EXPECTED VALUE JUDGMENTS

I have now argued from several angles (the Probability Dysfunction theory, risk concentration, and priority of prospects) that bioprediction of risk of harm is relevant to the PV. I have also showed via the theory of well-being and reliable agency the information to be relevant regardless of whether one knows one's own risk. From this discussion, we might arrive at the probability-weighted priority view.

> *Probability-Weighted Priority View (PWPV)*: it matters morally who we benefit; we should give priority to the worse off. Being worse off is a function of expected current well-being and expected future well-being.

In this model, a reduction in well-being to a person in the present could be outweighed in a comparison with a high risk of future harm in another.[13] Thus, someone with a bad stomach flu that is likely to resolve in two days could be better off than someone with a cough and high risk of conversion to pneumonia. We already have a mechanism that captures this idea: the shape of the probability curve developed in the reorientation of the concept of disorder in Chapter 4. This same careful risk banding can be used in thinking about priority setting.

Any practical decision, therefore, depends on expected-value judgments. When we judge that having schizophrenia, being poor, being tortured, not having access to education, living in an area with poor sanitation and high rates of tuberculosis, or even having tuberculosis makes someone worse off, we rely not just on a single expected value judgment, but a series of "nested" expected value judgments. One marker may show an increased risk of *x*, which is then more likely to result in *y*, which is then more likely to lead to *z*. For example, having the MAOA-L gene variant makes it more likely that, should one be abused as a child, one will develop a predisposition toward impulsive violence. This predisposition will in turn make it more likely that one will commit violent criminal acts, which will in turn make it more likely that one will be arrested and prosecuted, which will in turn make it more likely that one will be incarcerated and denied the opportunity to pursue intrinsic goods. At each level, the person's prospects change. At each level, we have multiple sets of factors that modify the likelihood of progression to the next level.

Although we have typically used the appearance of traditional symptoms (coughing, colored mucus, etc.) and disorder diagnosis to identify the worse off in the domain of health, this information is increasingly available from bioprediction. Predictive biomarkers hold the potential to show both that a person's prospects are worse than we thought and, conversely, that the prospects of others are better than we thought. Biomarkers, therefore, should be viewed as doubly important to priority setting according to the PWPV.

A COMMITMENT TO GIVE PRIORITY TO THE WORSE OFF REQUIRES THE DEVELOPMENT AND PROVISION OF BIOPREDICTIVE TECHNOLOGIES

If we operate under the PWPV, and so think it is important to give priority to the worse off, then we gain an obligation to identify the worse off. How do we currently know who is worse off? If we think it would be callous of institutions to distribute without trying to gain information on who is worse off, then how do we decide how much (and the type of) information they should gain (or even would be blameworthy for not gaining)? I argue here that it should not matter if worse prospects are estimated via traditional or nontraditional means (bioprediction). Consider the following three cases:

Shrapnel: The Bald-Eagle infantry unit is to be deployed to the front lines of battle in a high-shrapnel area. As a result, Bald-Eagle is at 50 percent risk of developing incapacitating emotional disturbance due to brain damage.

Mercury: Quicksilver unit was stationed on the western flank. It was recently discovered that the water source they happened to be using had been contaminated with high levels of mercury, a known neurotoxin. As a result, Quicksilver is at 50 percent risk of developing incapacitating emotional disturbance due to neuronal toxicity.

Biomarker: Within Delta unit, deployed on the front lines, a novel set of biomarkers identified that some of the unit is at 50 percent risk of developing incapacitating emotional disturbance due to the psychopathology of post-traumatic stress disorder (PTSD).[14]

For purposes of priority, it does not matter whether the group is worse off because they are in a high shrapnel area, because their water supply is contaminated, or because they possess a predictive biomarker. These different markers very well may be important to help decide what sort of resources to distribute, but the magnitude and the type of risk are what matters in terms of priority, and each, here, has an equal claim. If we have a sum of money that can be used to better the prospects of some, therefore, priority should be given to Bald-Eagle, Quicksilver, and the high-risk portion of Delta units over other units.

At the moment, we are cognizant of worse prospects due to the cases in Bald-Eagle and Quicksilver units but not of the high risk portion of Delta unit. Without biomarkers of PTSD-risk, those with this type of worse prospects would systematically be excluded from priority-driven resource allocation processes and effectively made even worse off. So if we accept a claim that social institutions have some sort of obligation to avoid making the worse off even worse off, the *development* of such biopredictive tests should therefore itself be a priority.

The situation is also unacceptable if biopredictive tests for high risk of brain-based disorder are developed, but not provided or accessible. To see why, consider the following example, *the PTSD Bioprediction Certificate*, which illustrates one possible consequence of bioprediction accessible only to the wealthy.

The PTSD Bioprediction Certificate: During the Vietnam War, the United States instituted a military draft. While many Americans went to war willingly, others employed a variety of means to escape the draft. Some fled the country. Others tried to get desk jobs or other military positions that entailed higher survival rates. Imagine that tests for PTSD risk were available, but only to those who could afford them. Those with risk of PTSD of a high-risk band could have this result printed in a bioprediction certificate and could have used it to argue that their prospects for coming away from battle intact are much worse than others and they should therefore have priority to resources that compensate for that risk, such as safer or shorter deployments. Those wealthy people at a low risk band would simply withhold

> the information. Poor conscripts, however, would not be able to afford the test. Consequentially, those poor conscripts at high-risk bands for PTSD would not be recognized as such and would become even further worse off than their wealthy counterparts, who experienced priority compensation.

The situation of the bioprediction certificate doubly penalizes those who are already worse off, a sort of "double jeopardy"; in order to avoid this wrong, we could require that biopredictive tests be provided. Only if biopredictive tests were provided could institutions be guaranteed the least distorted picture of who the worse off are.

NOT ONLY BY TREATMENT: MORAL USEFULNESS AND AN OBLIGATION TO DEVELOP BIOMARKERS

As the development of bioprediction continues, we are likely to have a mismatch between prediction and prevention. Specifically, we may identify high-risk categories of x, but not know how to use medicine or technology to decrease the risk of x. Some may argue that bioprediction is relevant to priority-setting and justice only when we have a clinical treatment to reduce the risk of harm.

There are many ways outside the clinics to modify the relationship between the risk of brain-based disorder and the risk of *harm*, however, and thus benefit a person *even if one can do nothing about the brain disorder directly though medicine*. It is seldom true that there is nothing that can be done to improve prospects, even of a medically untreatable condition. To illustrate this principle abstractly, consider that even if one may not be able to decrease the risk of rain on Sunday, one can decrease the risk that it will "ruin your day" by creating plans for Sunday that are more resilient to the weather. To illustrate the principle concretely, consider that better risk estimation of Alzheimer's disease could be *morally useful* to a person by enabling her to better plan financially for the contingencies more accurately weighted than by one's intuitions about what the probabilities might be; while enabling such financial planning would do nothing to decrease the *risk of dementia*, it could certainly do something to reduce the risk of harm to herself and her loved ones from inaccurate financial planning. The conceptual framework we established with the Probability Dysfunction concept should help us be precise on what we are defining as harms, which harms we are predicting, and which might be amenable to being reduced through bioprediction itself.

Where biopredictive information would be useful only with the provision of other resources, moreover, there may be an obligation of social institutions to make those resources more accessible. As I showed in Chapter 6, the provision of good public transportation might benefit someone at high risk of seizure (so they can pursue goods without risking harm to others), even if one can do nothing about the risk of seizure. Though free public transportation would benefit everyone, few cities could support such a system; providing free or discounted transportation to a person at high risk of seizure, however, might be both feasible and effective.

As I alluded to in Chapter 7, moreover, since knowledge of the level of one's own risk of harm to others can be relevant to one's level of responsibility, there might be situations where we might blame institutions that possess but withhold or have the ability to create that knowledge. Those in the *New York Times* story described a feeling that somehow G.M. is blameworthy for knowing that a certain make and model of car had a risk of power failure but not informing potentially affected individuals, even in cases where crashes had already happened and thus prevention was impossible. Where the information from bioprediction is relevant to legal responsibility, therefore, knowledge of the risk of harm due to imprisonment associated with a certain predisposition can be informative by enabling reduced punishment.

The claims to priority do not disappear, therefore, if there is no medical treatment capable of reducing a particular high-risk state. In sum, the distribution of social goods in response to those worse off in health need not be restricted to the medical domain. A social obligation to conduct further research may be suggested by these cases, though a thorough exploration and defense is beyond the scope of this book.

Buchannan et al. speak of the "colonization of the natural by the just" as a mark of moral progress:

> [W]hat we have taken to be moral progress has often consisted in pushing back the frontiers of the natural, in bringing within the sphere of social control, and thereby within the domain of justice, what was previously regarded as the natural, as merely a matter of good or ill fortune [. . . sometimes] we actually extend the social by expanding our control (e.g., the annual flooding of a great river is transformed from a fact of nature to an object of administration). (Buchanan et al. 2000, 83)

The knowledge provided by bioprediction of a phenomenon that we cannot control is similar to predicting the flood of the great river. Although we may be unable to halt the phenomenon, we can almost always minimize the damage (by evacuating the city around flood time). At least part of what makes the flooding of the river bad is that it is unpredictable. So, too, can we sever some of the contingent links between harm and disorder with better predictive information. If we cannot halt a seizure, better prediction can help us ensure that the person is in a safe environment with good access to medical care when a seizure occurs, and provide him with the resources to minimize the effect of his seizure predisposition on well-being.

STEPPING BACK TO PRIORITY

In this chapter, I have argued that biopredictive tests are capable of identifying groups of people who are worse off due to reduced prospects for welfare, even if the people themselves do not know their risk. I have pointed out the value of bioprediction to a range of distributive theories that have an overlapping consensus that institutions should give priority to the worse off. I have argued that bioprediction, in order to find and give priority to the worse off, should be developed and provided. The strengths of these arguments depend partly on the moral theory adopted. While the claims are strongest for those who wish to maximize value, these arguments apply to egalitarians and prioritarians alike. Because we live in a pluralistic society and our institutions do not follow one moral theory exclusively, I have focused on the overlapping consensus of giving priority to the worse off to make my arguments relevant to as great a number of people as possible.

Frank Jackson, in his seminal paper, "Decision-Theoretic Consequentialism and the Nearest and Dearest Objection," justifies a preference to giving benefits to one's own family partially because one generally has better knowledge of the expected value of that benefit (Jackson 1991). In the past, one could have made a similar argument for giving priority to those with traditionally defined disorders over prevention; we were more certain about the prospects of those with traditionally defined disorders, and thus could be more confident that we were helping the worse off. With the development of biomarkers, this justification will become unavailable.

NOTES

1. For more information about the Family Van, see www.familyvan.org.
2. There are, of course, many kinds of egalitarians; see Dworkin on equality of welfare (Dworkin 1981a) and resources (Dworkin 1981b). Amartya Sen and Martha Nussbaum speak of equality of *capabilities*, which, as Buchanan and colleagues summarize, is "not simply about the resources we have available to us, but our ability to transform those resources into actual capabilities, since variations in individuals will lead to different capabilities for the same inputs of resources" (Buchanan et al. 2000, 131). Luck egalitarians argue that we should be concerned only with inequality that is not the result of free choice (Arneson 2000). Luck egalitarianism has obvious connections with bioprediction, given my treatment of responsibility, but a proper investigation will be another project.
3. She could, like Bernard Williams (1973), claim that there is a moral disvalue in being the author of harm that outweighs the value of the greater reduction in inequality. However, in order to make a claim that it would be preferable to benefit

the worse off, even if benefiting a middle group would lead to a greater reduction in inequality, Oriol would need to look to the priority view. This chapter is primarily concerned with the moral effects of the identity of those benefited.

4. A different threshold view is Norman Daniels and colleagues's "level playing field," based on species-typical normal functioning (Buchanan et al. 2000, 141–144). Justice demands equality up to a point (the leveling of the playing field), beyond which inequalities may be acceptable. It is a demand of justice to raise the disadvantaged to a level of normal functioning (in health), but no obligation to raise them beyond that, even if they are still worse off compared to the best functioning.
5. For simplicity, we should assume that benefiting the better off does not also benefit the worse off.
6. If we grant this assumption, proponents of the PV might be subject to a kind of *distributive luck*.
7. Some forms of priority views would argue that identifying the worse off is only important if the weighted value added of benefiting the worse off outweighs the value of the benefit that would result from the direct use of the funds that are necessary to identify the worse off. Others, the egalitarians in particular, would view inequalities in health as an injustice, which could be magnified by an increase in such inequalities due to the ignorant distribution of goods.
8. Daniels, for example, has argued that differences in health should be given particular importance in distribution (Buchanan et al. 2000). I will not argue the centrality of health here, but I note that factors that negatively impact health are relevant to those who endorse the PV.
9. Daniels tries to resolve the disagreement by considering a 100-pill version, where 100 people have been exposed to Cory, and each have a 1 percent chance of contracting the disease. "It now seems even less plausible to reduce the one in one hundred chance of contagion and death for each person rather than save the one person already infected" (Daniels 2012b). This 100-pill version, I think, fails because it does not correct for the risk of multiple people dying or for risk aversion when there is a certain death versus uncertain ones.
10. Decision theorists may multiply the probability of each outcome by the moral value of that outcome; only if there is a direct linear transformation between the number of lives lost and the disvalue of that outcome (and a value of zero assigned to the case where none of the five dies) will the decision theorist agree that the scales are balanced when it is finally time to consider risk concentration.
11. Daniels and colleagues's level-playing-field conception of equal opportunity naturally incorporates the idea of uncertain prospects (Buchanan et al. 2000, 141–144).
12. Theories of well-being are introduced in Parfit (1984, 493–502).
13. Of course, the time of death does seem to matter. If Bob has an 80 percent risk of dying over the next 5 years at age 20 and Agnes has an 80 percent risk of dying over the next 5 years at age 80, Bob is worse off than Agnes. It is this intuition that underlies the phrase "untimely death," and the fair-innings arguments for giving healthcare priority to the sick young; see Hope et al. (2008).
14. Orr et al. (2012) illustrates emerging efforts to predict vulnerabilities to post-traumatic stress symptoms. Pitman, an author on that paper, is part of a consortium that is undertaking a massive trial to identify predictive biomarkers of PTSD; see Drummond (2012).

Conclusion

The potential benefits of applying biomarkers to medicine and other institutions are important forces driving the development of biomarkers; few have considered, however, the potential benefits of the development of biomarkers for the.philosophical foundations of these institutions.

In this book, I have considered philosophical and ethical issues arising from the development of bioprediction in psychiatry and neurology. I have outlined the motivation behind the development of predictive biomarkers, and have described some properties of the type of information they can provide. Namely, these technologies greatly expand the domain of health in which we can quantify the probabilities of a future adverse event, from future seizure or psychosis to dementia. The proliferation of quantified, probabilistic information about health risks is challenging a long-embraced social and philosophical convention: that disorder is a binary or categorical concept. I have argued that much of the ethical controversy surrounding the development of bioprediction actually stems from efforts to preserve this binary conception of disorder, but it is a heuristic that is no longer tenable.

In Chapter 3, I invited Nasrudin and *King Lear*'s Cordelia to show us the dangers of reifying existing categories of disorder in mental health. I argued that in light of emerging research in the neurosciences, which suggests both extensive heterogeneity within and overlap between existing categories, we should reorient those categories around emerging biomarkers. Such a reorientation requires a probabilistic concept of disorder to successfully capture the pleiotropy that exists at the biological level. In Chapter 4, I proposed the Probability Dysfunction model of disorder to both capture this pleiotropy and enable a synergy with probabilistic information about health. I showed increased probability of future harm to be an integral part of any model of disorder: orienting the concept around risk of harm uniquely enables us to deal with chronic but managed conditions, periodic conditions (e.g., epilepsy), conditions with signs and symptoms that themselves do not bring present harm but greatly increase the risk of harm in the future (e.g., hypertrophic cardiomyopathy), and prognoses. I raised concerns that the proposed revision to the definition of "disorder" found in the *DSM-5*, because it removes the clause about increased risk of future harm, consequently risks exacerbating confusion about probability. I explicitly oriented the Probability Dysfunction model, in contrast, with risk of future harm at its core. The Probability Dysfunction model avoids the problems of three existing models of disorder from the philosophy of medicine by remaining agnostic about natural teleology and making explicit debate about whether increased risk of harm is due to injustice. Although risk of harm is a continuous function, in Chapter 5 I demonstrated the benefit of stratifying individual disorders based upon risk magnitude and type of harm, a concept I named "risk banding," and matching response to risk bands. I showed that applying risk banding allows one to avoid many of the ethical concerns raised about including categories based on risk of psychosis or risk of dementia in psychiatry and neurology. I thereby constructed a practical framework with which to consider the expanded information of risk brought by predictive biomarkers.

In Part II, I considered the impact of the development of bioprediction on individual responsibility. I began in Chapter 6 by showing how bioprediction of adverse involuntary events related to brain function enables a degree of enhanced remote control over the risk that those events pose to others. This enhanced remote control seems sufficient to create obligations to protect others from risk of harm. As a model for the degree of foreseeability required for moral responsibility and the degree of precaution warranted, I considered the risk of seizure while driving. I showed that the per-trip probability of a seizure that leads to harm of others is very low and that the obligation to protect others from that risk deemed reasonable by current legal structures is very strong. I argued that applying these thresholds to the risks likely to be revealed by bioprediction will significantly expand the domain of actions that pose sufficient risk to others for an agent to be held morally responsible. I argued that in many cases, the failure to gain information from bioprediction fits the criteria for a benighting act, and so turns unknowingly risking harm to others into culpable ignorance. I also raised the possibility that bioprediction creates obligations for policymakers, scientists, doctors, and others who know about the technology to inform others.

When the technology exists, but individuals are uninformed, I examined the possibility that biopredictive technologies could also rightly diminish moral responsibility. I considered as a case study the MAOA gene x environment interaction, which is thought to increase the gravity of provocation perceived by the individual, and thereby increase the likelihood of an impulsive violent reaction. I pointed out that courts, in practice, already depend heavily on probabilistic evidence, for example that a given condition or set of circumstances (such as a threat or a particular delusion) increases the likelihood of violent reaction via a morally relevant mechanism; like incorporating probabilistic information into disorder, therefore, there is no incompatibility with incorporating probabilistic biological predispositions into the legal context; doing so, however, requires recognition that binary conceptions of the relevance of disorder to moral responsibility have been heuristic, and we should explicitly embrace probabilistic models.

In Part III, I engaged directly with some of the issues foreshadowed in the previous sections, specifically obligations of society to the individual. In Chapter 8, I raised the possibility that society has been setting health-related priority in ignorance of an important way of being worse off, namely being at increased risk of harm. Because bioprediction is precisely oriented toward estimating risk of future harm, I argued that those who endorse giving priority to the worse off have a *prima facie* obligation to develop and provide access to biopredictive testing, even if there is no specific medical treatment for the risk predicted. In constructing these arguments, I engaged with several objections to the moral relevance of bioprediction to distributive justice, which I showed to largely be untenable objections to the relevance of probabilities and risk. For practical purposes, I applied the concept of risk banding to suggest that priority could be matched to risk band.

The overarching purpose of this book is an analysis of the negative results of categorical thinking about disorder in psychiatry and neurology in light of the development of bioprediction, and the potential benefits of a switch to probabilistic thinking. I have offered the Probability Dysfunction theory and risk banding as one model that can help us make this switch. This book, I am aware, raises many more questions than it answers. I will finish with brief consideration of two issues related to the book's overarching purpose.

First, one observation that has arisen in this book is that each of the areas mentioned (i.e., medicine, moral responsibility, and priority) is concerned with slightly different probabilities. Medicine is concerned with increased probability of harm to self, subtracting the increase due to injustice. The domain of moral responsibility is concerned with increased probability of harm to others; responsibility is enhanced if one knew or should have known about an increase in risk of harm to others and did not take proper precautions; responsibility is diminished if a morally relevant mechanism behind that increase in risk of harm is identifiable but unknown to the agent (e.g., increased emotional gravity of provocation to the defendant). Those who wish to give priority to the worse off are also concerned with increased probability of harm to self, but consider increases equally whether or not they are due to injustice.

A potential hazard for the domains of responsibility and priority arises from the dominance of the medical domain in directing and conducting research on predictive biomarkers. As a result, the purpose of such research is to identify tools that translate from the laboratory to the medical clinic and may lead to better clinical treatments. I spent some time in Chapter 7, however, outlining that there will often be very specific unanswered questions about the mechanisms behind increased risk that must be answered for bioprediction to be relevant in the courtroom. This set of questions will seldom match clinical research priorities. What structures are set up to incentivize this sort of morally or legally useful research? While the US government funds a series of medical scientist training programs designed to fill the gaps and meet the needs of practical clinical medicine through medically informed basic science research, there are no such governmental programs designed to do the same for law. Perhaps it is time to consider a "legal scientist training program" to fill the gaps and meet the needs of the courtroom in regard to bioprediction, as I proposed at the end of that chapter.

Finally, making the switch to the probabilistic thinking necessary to successfully address the looming ethical issues raised by the development of bioprediction will require significant changes in the education of doctors, medical researchers, lawyers, policymakers, and the general population. The current level of training regarding thinking about risk and probability in each of these groups is inadequate to accommodate this conceptual shift. I made this point about the need to increase risk literacy in Chapter 4, but it is so essential that I feel it is only appropriate that it conclude this book.

APPENDIX I

A Brief Note on Genetic versus Non-Genetic Biomarkers

Many countries extend certain protections to people against the disadvantageous use of predictive genetic information (DNA, RNA, or family history), especially by insurance companies. With the advance of proteomics, lipidomics, medical imaging, and the army of other biomarkers, however, the extent to which it is becoming possible to estimate risk of future illness (the diagnoses themselves serving as surrogate endpoints for other harms) via non-genetic biological means is on the verge of exploding. Because these non-genetic predictive biomarkers, as a class and through multiplexing, show potential for much higher predictive validity than their genetic counterparts, they prompt us to ask the question: Should we extend non-discrimination policies to all forms of bioprediction? And if not, what are the morally relevant differences between predictive genetic and non-genetic biomarkers that justify such different national policies? A rigorous response to these questions is beyond the scope of this appendix. I will, however, raise the following potential differences in predictive power, family privacy, identity, and brute luck, though I think they fall short of justifying such exceptionalism for genetic biomarkers in social policy.

PREDICTIVE POWER

A first argument one might appeal to is that the predictive value of genetic biomarkers is usually either very low (common variants of small effect) or very high (the dominant mutations in the Huntington's gene). Because the high predictive value variants are rare but well known, one might worry that there is a greater potential that the common variants will be *overweighted* in an insurance company's risk calculus, and that this would be unfair. Because non-genetic biomarkers span the whole intermediate range of predictive power, one might think it less likely that these biomarkers would be over-weighted. I think that this argument falls short for two reasons. The first is that one might reasonably imagine that quantitatively oriented insurance companies would be the *least likely to overweight*, and second, that even if insurance accurately captured the risk information in genetic variants, proponents of policies like GINA would be unlikely to think that using genetic information is acceptable.

PRIVACY OF FAMILY

A second argument one may appeal to is to observe that genetic risk testing in one person is likely to produce information on the risk status of other family members who also may possess the same variants, and this shared information is reason to treat genetic biomarkers differently from non-genetic ones. Non-genetic biomarkers associated with heritable conditions, however, would also provide the same sort of shared risk information.

IDENTITY AND PRIVACY

A third argument might be to posit that genetic information requires special protections because it is more intimately tied to identity; this is the "you are your genes" argument. However, it is unclear as to why genes would be more integral to identity than your proteome, which more faithfully integrates your individual environmental influences, or for that matter, than the size and function of your hippocampus (a region of the brain very important in the consolidation of new memories and often disrupted in illness, from schizophrenia, to depression, to Alzheimer's disease).

BRUTE LUCK

A fourth, and I think strongest, argument takes a page out of luck egalitarianism. Luck egalitarians argue that we have social obligations to eliminate limitations in opportunity due to factors that are not the result of one's choices (Arneson 2000); this would include both the effects of the "social lottery" (whether the family you happened to be born into is wealthy or poor, etc.) as well as the "genetic lottery" (what genes you get). While it is clear that differential susceptibilities to disease due purely to genetic variation are not due to choice, the case is less clear with non-genetic biomarkers, as they integrate the effects of environment, which are sometimes a matter of choice, or "option luck." Luck egalitarians, therefore, may argue that non-genetic biomarkers do not raise the same social obligations as genetic ones. At a bare minimum, however, even luck egalitarians would be committed to minimizing the limitations due to biomarkers in children up until the age at which they can be considered responsible in the way that would disqualify the claim of the biomarker.

APPENDIX II

Seizure Prediction

Approximately 1–2 of 20 people will experience a seizure at some point in life (Bragin et al. 2010; Nestler et al. 2009). Particularly pertinent to our investigation here is the presentation of first seizures in those persons of driving age; this is the category that might feasibly fall into Probability's shoes. Of those presenting with first seizures in Australia, for example, 60 percent are above 16 years of age, the age at which persons are eligible to enter with supervision the driver's seat of a car; new onset epilepsy is the most common cause of first seizure with 50 percent of those with a first unprovoked seizure experiencing another in the next 8 years (Adams & Knowles 2007).

GENETICS

Although we are concerned with adolescent/adult onset events, it is worth starting off by mentioning that about 10 percent of early-onset forms of absence epilepsy are attributable to mutations in the gene encoding glucose transporter type 1, *SLC2A1* (Pal et al. 2010). The genetic components of other epilepsies are less well categorized, and it is likely that most alter seizure thresholds.[1]

What is reasonably clear, however, is that interfering with the proper expression of a range of neuronal-excitability-related genes can lead to seizures and epilepsy. More than 20 strains of knockout mice, for example, develop spontaneous seizures[2] (Nestler et al. 2009). Most of these mutant strains exhibit seizure onset before mid-adolescence (and are thus less interesting to us here), but several mutants exhibit late-adolescent to adult onset: Weaver (g-protein gated inward rectifier girk2) has adult onset; Centromere bp-b (brain specific DNA binding protein) has onset at 3–4 months; Synapsin 1,2 has onset after 2 months. In these mice raised in standard laboratory conditions, the genotype can predict seizure onset. No human knockout of this type has been described. There is, however, a possibility that future research will describe risk variants in humans.

Along these lines, and perhaps more relevant in our case, there exist a large number of mouse mutants that exhibit decreased seizure threshold and susceptibility: these are in the signaling proteins, Tyn tyrosine kinase receptor, tPA, and BDNF. It is thought that the absence of these proteins upsets the balance of excitation and inhibition in key areas of the brain, leading to seizure susceptibility (Nestler et al. 2009). A decreased seizure threshold becomes important when we remember that we bathe our brains in insults that at any given moment may tip the scales of excitation and inhibition; if the scales tip too far by this chaotic process, a seizure might ensue. In other words, differences in seizure susceptibilities at population levels will translate to differences in seizure incidence.

To highlight even more directly the combinatory requirement of biology and environment, we can consider the following cases. One mutation called γ2R43Q in the GABAa receptor, a major inhibitory receptor in the brain, decreases the seizure threshold in adult mice; because even temporary changes in GABAa receptors early in life can have lasting impact on seizure threshold in adulthood (Chiu et al. 2008) and because temperature increase impairs this particular mutant receptor's activity (Galanopoulou 2010), it raises the possibility that childhood fever might permanently alter the seizure threshold in those individuals who possess this variant. Mice with a mutation in the nicotinic acetylcholine receptor (L247T) display a higher sensitivity specifically to nicotine-induced seizures than mice without the mutation (Steinlein & Bertrand 2010). Finally, head injury in rodents leads to differential seizure-threshold lowering depending on genetic background (Pitkänen et al. 2011), which suggests that the rodents' specific genetic makeup can provide resilience against or susceptibility to environmental insults.

Because the electrical activity of a seizure seems to depend on the balance of excitation and inhibition across key circuits, much attention has recently been paid to the proteins most closely tied to these electrical properties, namely, ion channels. Aside from the GABA channels noted earlier, seizures have been associated with mutations in the major glutamate channels (NMDA and AMPA) and with potassium channels (Endele et al. 2010; Klassen et al. 2011; Olson & Terzic 2010). Further corroborating the channel hypothesis, Rasmussen encephalitis, a disorder in which auto-antibodies bind to and interfere with glur3 subunit of the AMPA receptor, can cause seizure onset usually at 2–10 years of age in humans (Nestler et al. 2009).

One group of researchers recently endeavored to construct a "channotype," a profile of variation in all known ion channel genes (Klassen et al. 2011). It was hoped that the channotype would be a barometer that would enable individualized epilepsy onset predictions. The group sequenced 237 channel genes in 139 healthy humans and 152 with idiopathic epilepsy (i.e., epilepsy with no known cause). Rather than a signature of epilepsy, the group found a variety of polymorphisms that did not significantly predict individual risk; even healthy people could have enrichment of deleterious mutations. The authors conclude that "personalized ion channel disease prediction will require the integration of genomic profiles with single-cell proteomics and computational modeling of dynamic circuit behavior to translate unique channel variant portfolios into more informative predictors of personal risk for excitability disorders." In other words, this goal might be difficult. One critique is that this study is cross-sectional (done at one point in time), so it is unclear whether any of these healthy controls with the deleterious mutations might develop seizures in the future. The other critique is the environmental one outlined earlier: that perhaps a channotype combined with an outline of key environmental exposures could identify subpopulations with identifiable predictive risk.

EVENT RELATED

Immature rodents exposed to viral RNA or LPS (a component of the bacterial capsule) show lower seizure threshold as adults, an effect that seems mediated by pro-inflammatory immune response that triggers long-term glutamate receptor subunit changes in the hippocampus (Ravizza et al. 2011)—all effects that could, in theory, be monitored and entered into an actuarial risk calculator.

After stroke, 2–4 percent of patients develop recurring seizures (compared to 1 percent of the general population) (Kwan 2010). After head trauma, up to 25 percent develop seizure; even after mild head trauma (i.e., concussion), some patients exhibit residual sub-threshold epileptiform activity 12–14 months after the incident (Pitkänen et al. 2011).

Interestingly, 5-year risk for first seizure was 23 percent with arteriovenous malformations with intracranial hemorrhage or focal neurological deficit ($n = 119$; CI 9–37) (those that did not present with seizure; all those presenting with seizures had further seizures, as pointed out in Hauser & Mohr [2011]; Josephson et al. [2011]).

TRANSCRIPTOMICS

As all of the previously noted event mediators—viral infection, stroke, head trauma, and hemorrhage—have inherent immune system components, and as emerging research in the field of neuroimmunology is beginning to show that the neural and immune systems are much more intimately associated than once thought, one group recently endeavored to assay the transcriptome of peripheral blood cells in animals for association with later epilepsy. The authours summarize their results:

> The obtained microarray data point to distinct groups of genes associated with animals that later develop seizures. This supports the hypothesis that the molecular signature preceding the development of epilepsy is present in the peripheral blood transcriptome, and this in turn may allow the development of a prognostic test that can be used to both screen and diagnose potential epilepsy patients, and to prospectively evaluate the effectiveness of ant epileptogenic therapy. However, more work is required in order to fully demonstrate the usefulness of this approach. (Karsten et al. 2011, p. 215)

ELECTROPHYSIOLOGICAL

Pathological high-frequency oscillations (pHFOs), which from animal studies are thought to reflect fast and synchronized activity across populations of neurons, might reflect the neuronal abnormalities responsible for epilepsy. Normal HFOs are 100 Hz, whereas pathological ones are 250–600 Hz and are only present in the dentate gyrus[3] of rats that later exhibit spontaneous seizures (Bragin et al. 2010). Though HFOs have been detected in humans, their successful differentiation from pHFOs is an ongoing project. One difficulty is that the pHFOs in rats are detected with implanted electrodes in the brain, whereas we might seek to avoid this option in humans.

Another strategy is to train an algorithm over time to pick out patterns in neural activity that tend to precede seizure in a particular individual. In 2013, Cook and colleagues reported the results of a proof of principle study in which algorithms were trained with electrophysiological data recorded from electrodes implanted in the brains of people with epilepsy in order to predict short time periods of low, medium, and high risk of seizure (Cook et al. 2013). Though this study showed variability in the quality of predictions person to person, it holds the potential for a personal and temporally restricted risk prediction of just the sort that I argued in Chapter 6 could provide a "solution from technology" to the problem of obligations to others becoming more burdensome as the number of forseeable biological risks we pose to others increases.

SUM AND COMBINATIONS

There may be even greater potential in these technologies through their combination. If we combine an already high risk category, like the patients presenting with arteriovenous malformations, with genomics, transcriptomics, and electrophysiological assays, the possibility of identifying very high-risk groups may soon exit from science fiction.

NOTES

1. A seizure threshold, approximately speaking, might be considered to be set at the intensity level of an insult (audio, visual, chemical, thermal) necessary to trigger a seizure.
2. A knockout mouse is a strain in which a specific gene has been genetically manipulated by an experimenter to be non-functional (i.e., it is as if the gene has been "knocked out of the genome").
3. The dentate gyrus is a brain area thought to be implicated in epilepsy because of its integral inhibitory function to the hyper-excitable hippocampus.

APPENDIX III

The Case of Stefania Albertani

This appendix is excerpted from Baum and Savulescu (2014, pp. 25–26).

> In the court of Como, Italy, Stefania Albertani was tried in 2011 for the murder of her sister and the attempted murder of her parents (*Albertani* Gip di Como, 20.08.2011; Feresin 2011). The sentence was reduced from 30 to 20 years (3 years of which were to be spent in a mental institution) by the weight of a combination of traditional psychiatric analysis, structural brain imaging, memory tests, and genetic testing, which showed that Albertani possessed the MAOA-L, as well as two other genes weakly evidenced to be associated with impulsive violence.
>
> Albertani had administered benzodiazepine sedatives to her sister over a period of several days, which caused her sister to be in a confused state. Neighbors called the police and the sister was taken to the hospital, but Albertani convinced the sister to return home without consulting the doctors. Albertani then murdered the sister and burned her body in the backyard, after first telling the neighbors she was about to burn some paper so that they would neither be worried by the smoke nor call the police. Albertani then took identification and credit cards off the body and used them for some time. During this time, she also created threatening letters, impersonating her sister, and sent them to various family members. On a separate incident, Albertani had attempted to set fire to her parents' car with them inside. She had also on another occasion given her father a harmful dose of medication.
>
> Albertani was arrested, however, upon the attempted murder of her mother. After the disappearance of the sister, the police had "bugged" the parents' house and entered the scene after hearing screams and sounds of struggle. They found the mother partially unconscious with her clothes on fire and were able to remove her to a hospital.

There were initially two expert psychiatric testimonies, but they came to opposite conclusions. As such, the judge permitted the defense to bring in a third expert testimonial, during which neuroimaging, behavioral, and genetic testing were done. Besides the genetic polymorphisms already mentioned, Albertani was shown to have abnormalities in the anterior cingulate gyrus and the insula as compared with 10 female controls (Feresin 2011). [. . .]. The experts also used behavioral tests to estimate dissociation from memories (as Albertani claimed not to remember attacking her mother), and by the combination of these measures assigned Albertani a dissociative identity disorder.[1]

NOTE

1. A disorder characterized by the alternation of (often subtly) distinct personalities within the same person, accompanied by amnesia of the events and actions of the other personality; often accompanied by childhood trauma, and hypothesized to be a defense mechanism.

REFERENCES

Adams, S. M., and Knowles, P.D., 2007. Evaluation of a first seizure. *American Family Physician*, **75**(9), pp. 1342–1348.

A-G for Jersey v. Holley [2005] UKPC 23; [2005] Crim LR 966.

Ahuja, A., 2009. The get out of jail free gene. *The Sunday Times (UK)* (November 17).

Albertani, Gip di Como, 20.08.2011. Giudice per le le indagini preliminary di Como, Maria Luisa Lo Gatto decisione del 20 agosto 2011. Full transcript http://static.ilsole24ore.com/DocStore/Professionisti/AltraDocumentazione/body/12600001-12700000/12693249.pdf. In: Machiochhi, P. *Gip di Como: le neuroscienze entrano e vincono in tribunale.* http://www.diritto24.ilsole24ore.com/guidaAlDiritto/penale/primiPiani/2011/08/le-neuroscienze-entrano-e-vincono-in-tribunale.html

Alivisatos, A. P., Chun, M., Church, G. M., Greenspan, R. J., Roukes, M. L., and Yuste, R., 2012. The brain activity map project and the challenge of functional connectomics. *Neuron*, **74**(6), pp. 970–974.

American Psychiatric Association (APA), 2000. *Diagnostic and Statistical Manual of Mental Disorders: DSM-IV-TR*. 4th edition. Washington, DC: American Psychiatric Association.

American Psychiatric Association (APA), 2013. *Diagnostic and Statistical Manual of Mental Disorders: DSM-5*. 5th edition. Arlington, VA: American Psychiatric Association.

Amminger, G. P., and McGorry, P. D., 2012. Update on omega-3 polyunsaturated fatty acids in early-stage psychotic disorders. *Neuropsychopharmacology*, **37**(1), pp. 309–310.

Anderson, K., Anderson, L. E., and Glanze, W. D., 1998. *Mosby's Medical Dictionary.* St. Louis: Mosby.

Aristotle, 1999. Book III. *Nicomachean Ethics.* 2nd edition, Terrence Irwin ed. Indianapolis, IN: Hackett, pp. 30–49.

Arneson, R. J., 2000. Luck egalitarianism and prioritarianism. *Ethics*, **110**(2), pp. 339–349.

Atkinson A. J., Colburn, W. A., Degruttola, V. G., Demets, D. L., Downing, G. J., Hoth, D. F., Oates, J. A., Peck, C. C., Schooley, R. T., Spilker, B. A., Woodcock, J., and Zeger, S. L., Biomarkers Definitions Working Group, 2001. Biomarkers and surrogate endpoints: Preferred definitions and conceptual framework. *Clinical Pharmacology and Therapeutics*, **69**(3), pp. 89–95.

Bahn, S., Noll, R., Barnes, A., Schwarz, E., and Guest, P.C., 2011. Challenges of introducing new biomarker products for neuropsychiatric disorders into the market. *International Review of Neurobiology*, **101**, pp. 299–327.

Barry, H., Hardiman, O., Healy, D. G., Keogan, M., Moroney, J., Molnar, P. P., Cotter, D. R., and Murphy, K. C., 2011. Anti-NMDA receptor encephalitis: An important differential diagnosis in psychosis. *British Journal of Psychiatry*, **199**(6), pp. 508–509.

Baum, M. L., 2011 (first published online). The monoamine oxidase A (MAOA) genetic predisposition to impulsive violence: Is it relevant to criminal trials? *Neuroethics*, doi: 10.1007/s12152-011-9108-6, 2013, **6**(2), pp. 287–306.

Baum, M. L., and Savulescu, J., 2014. Behavioural biomarkers: What are they good for? In: I. Singh, W. Sinott-Armstrong, and J. Savulescu, eds., *BioPrediction of Bad Behavior: Scientific, Legal, and Ethical Challenges*. New York; Oxford: Oxford University Press, pp. 12–41.

Bayout v. Francesco. [2009] RGAssise App. 6/2008 RGNR 1685/2007, RG. sent 5, dd 18 settembre 2009.

Beddington, J., Cooper, C. L., Field, J., Goswami, U., Huppert, F. A., Jenkins, R., Jones, H. S., Kirkwood, T. B. L., Sahakian, B. J., and Thomas, S. M., 2008. The mental wealth of nations. *Nature*, **455**(7216), pp. 1057–1060.

Beers, C. W., [1908]1981. *A Mind That Found Itself: An Autobiography*. 5th edition. Pittsburgh, PA: University of Pittsburgh Press.

Bloom, D. E., Cafiero, E. T., Jane-Llopis, E., Abrahoms-Gessel, S., Bloom, L. R., Fathima, S., Feigl, A. B., Gaziano, T., Mowafi, M., Pandya, A., Prettner, K., Rosenberg, L., Seligman, B., Stein, A. Z., and Weinstein, C., 2011., *The Global Economic Burden of Noncommunicable Diseases*. Geneva: World Economic Forum.

Bloss, C. S., Jeste, D. V., and Schork, N. J., 2011. Genomics for disease treatment and prevention. *Psychiatric Clinics of North America*, **34**(1), pp. 147–166.

Boenink, M., 2009. Tensions and opportunities in convergence: Shifting concepts of disease in emerging molecular medicine. *NanoEthics*, **3**(3), pp. 243–255.

Bonnett, L. J., Tudur-Smith, C., Williamson, P. R., and Marson, A. G., 2010. Risk of recurrence after a first seizure and implications for driving: Further analysis of the multicentre study of early epilepsy and single seizures. *BMJ (Online)*, **341** (c6477), doi: 10.1136/bmj.c6477.

Boorse, C., 1975. On the distinction between disease and illness. *Philosophy & Public Affairs*, **5**(1), pp. 49–68.

Bragin, A., Engel, J. and Staba, R. J., 2010. High-frequency oscillations in epileptic brain. *Current Opinion in Neurology*, **23**(2), pp. 151–156.

Brennand, K. J., Simone, A., Jou, J., Gelboin-Burkhart, C., Tran, N., Sangar, S., Li, Y., Mu, Y., Chen, G., Yu, D., Mccarthy, S., Sebat, J., and Gage, F. H., 2011. Modeling schizophrenia using human induced pluripotent stem cells. *Nature*, **473**(7346), pp. 221–225.

Buchanan, A. E., Brock, D. W., Daniels, N., and Wikler, D., 2000. *From Chance to Choice: Genetics and Justice*. Cambridge; New York: Cambridge University Press.

Buckholtz, J. W., and Meyer-Lindenberg, A., 2008. MAOA and the neurogenetic architecture of human aggression. *Trends in Neurosciences*, **31**(3), pp. 120–129.

Buckholtz, J. W., and Meyer-Lindenberg, A., 2014. MAOA and the bioprediction of antisocial behavior: Science fact and science fiction. In: I. Singh, W.

P. Sinnott-Armstrong, and J. Savulescu, eds., *Bioprediction, Biomarkers, and Bad Behavior: Scientific, Legal, and Ethical Challenges*. New York; Oxford: Oxford University Press, pp. 131–152.

Buxbaum, J. D., and Sklar, P., 2011. Human induced pluripotent stem cells: A new model for schizophrenia? *Cell Stem Cell*, **8**(5), pp. 461–462.

Campbell [1997] 1 Cr App Rep 199, [1997] Crim LR 227, CA.

Campbell, C., and Eastman, N., 2014. The limits of legal use of neuroscience. In: I. Singh, W. P. Sinnott-Armstrong, and J. Savulescu, eds., *Bioprediction, Biomarkers, and Bad Behavior: Scientific, Legal, and Ethical Challenges*. New York; Oxford: Oxford University Press, pp. 91–117.

Camplin [1978] AC 705, [1978] 2 WLR 679, 67 Cr App Rep 14, [1978] 2 All ER 168, HL.

Cannon, T., Chung, Y., He, G., Daqiang, S., Jacobson, A., Van Erp, T., Mcewen, S., Addington, J., Bearden, C., Cadenhead, K., Cornblatt, B., Mathalon, D., Mcglashan, T., Perkins, D., Jeffries, C., Seidman, L., Tsuang, M., Walker, E., Woods, S. and Heinssen, R., on behalf of the North American Prodrome Longitudinal Study Consortium, 2015. Progressive reduction in cortical thickness as psychosis develops: A multisite longitudinal neuroimaging study of youth at elevated clinical risk. *Biological Psychiatry*, **77**, pp. 147–157.

Caspi, A., Mccray, J., Moffitt, T. E., Mill, J., Martin, J., Craig, I. W., Taylor, A., and Poulton, R., 2002. Role of genotype in the cycle of violence in maltreated children. *Science*, **297** (5582), pp. 851–854.

Centerwall, S. A., and Centerwall, W. R., 2000. The discovery of phenylketonuria: The story of a young couple, two retarded children, and a scientist. *Pediatrics*, **105**(1), pp. 89–103.

Chiu, C., Reid, C. A., Tan, H. O., Davies, P. J., Single, F. N., Koukoulas, I., Berkovic, S. F., Tan, S., Sprengel, R., Jones, M. V., and Petrou, S., 2008. Developmental impact of a familial GABAA receptor epilepsy mutation. *Annals of Neurology*, **64**(3), pp. 284–293.

Chung, Y., and Cannon, T., 2015. Brain imaging during the transition from psychosis prodrome to schizophrenia. *The Journal of Nervous and Mental Disease*, **203**(5), pp. 336–341.

Coggon, J., 2008. On acts, omissions and responsibility. *Journal of Medical Ethics*, **34**(8), pp. 576–579.

Collins, P. Y., Patel, V., Joestl, S. S., March, D., Insel, T. R. and Daar, A. S., 2011. Grand challenges in global mental health. *Nature*, **475**(7354), pp. 27–30.

Cook, M. J., O'Brien, T. J., Berkovic, S. F., Murphy, M., Morokoff, A., Fabinyi, G., D'Souza, W., Yerra, R., Archer, J., Litewka, L., Hosking, S., Lightfoot, P., Ruedebusch, V., Sheffield, W. D., Snyder, D., Leyde, K., and Himes, D., 2013. Prediction of seizure likelihood with long-term, implanted seizure advisory system in patients with drug-resistant epilepsy: A first-in-man study. *Lancet Neurology*, **12**(6), pp. 563–571.

Cooper, R., 2002. Disease. *Studies in History and Philosophy of Science Part C: Studies in History and Philosophy of Biological and Biomedical Sciences*, **33**(2), pp. 263–282.

Corder, E. H., Saunders, A. M., Strittmatter, W. J., Schmechel, D. E., Gaskell, P. C., Small, G. W., Roses, A. D., Haines, J. L., and Pericak-Vance, M. A., 1993. Gene dose of apolipoprotein E type 4 allele and the risk of Alzheimer's disease in late onset families. *Science*, **261**(5123), pp. 921–923.

Crisp, R., 2003. Equality, Priority, and Compassion. *Ethics,* **113**(4), pp. 745–763.

Dalmau, J., Gleichman, A. J., Hughes, E. G., Rossi, J. E., Peng, X., Lai, M., Dessain, S. K., Rosenfeld, M. R., Balice-Gordon, R. and Lynch, D. R., 2008. Anti-NMDA-receptor encephalitis: Case series and analysis of the effects of antibodies. *The Lancet Neurology,* **7**(12), pp. 1091–1098.

Dalmau, J., Lancaster, E., Martinez-Hernandez, E., Rosenfeld, M. R., and Balice-Gordon, R., 2011. Clinical experience and laboratory investigations in patients with anti-NMDAR encephalitis. *The Lancet Neurology,* **10**(1), pp. 63–74.

Daniels, N., 2012a. Treatment and prevention. In: H. S. Faust and P. T. Menzel, eds., *Prevention vs. Treatment: What's the Right Balance?* New York; Oxford: Oxford University Press, pp. 176–192.

Daniels, N., 2012b. Reasonable disagreement about identified vs. statistical victims. *Hastings Center Report,* **42**(1), pp. 35–45.

Devlin, A., Odell, M., L. Charlton, J., and Koppel, S., 2012. Epilepsy and driving: Current status of research. *Epilepsy Research,* **102**(3), pp. 135–152. http://dx.doi.org/10.1016/j.eplepsyres.2012.08.003.

Dolmetsch, R., and Geschwind, D. H., 2011. The human brain in a dish: The promise of iPSC-derived neurons. *Cell,* **145**(6), pp. 831–834.

Drummond, K., 2012. Military PTSD: All-star scientists start mental health mega-project. *Forbes Business* (Sept 04), http://www.forbes.com/sites/katiedrummond/2012/09/04/ptsd-draper-labs/?ss=data-driven.

Dworkin, R., 1981a. What is equality? Part 1: Equality of welfare. *Philosophy & Public Affairs,* **10**(3), pp. 185–246.

Dworkin, R., 1981b. What is equality? Part 2: Equality of esources. *Philosophy & Public Affairs,* **10**(4), pp. 283–345.

Eastman, N., and Campbell, C., 2006. Science and society: Neuroscience and legal determination of criminal responsibility. *Nature Reviews Neuroscience,* **7**(4), pp. 311–318.

Endele, S., Rosenberger, G., Geider, K., Popp, B., Tamer, C., Stefanova, I., Milh, M., Kortüm, F., Fritsch, A., Pientka, F. K., Hellenbroich, Y., Kalscheuer, V. M., Kohlhase, J., Moog, U., Rappold, G., Rauch, A., Ropers, H.-., Von Spiczak, S., Tönnies, H., Villeneuve, N., Villard, L., Zabel, B., Zenker, M., Laube, B., Reis, A., Wieczorek, D., Van Maldergem, L. and Kutsche, K., 2010. Mutations in GRIN2A and GRIN2B encoding regulatory subunits of NMDA receptors cause variable neurodevelopmental phenotypes. *Nature Genetics,* **42**(11), pp. 1021–1026.

Eshleman, A., 2009. Moral responsibility. *The Stanford Encyclopedia of Philosophy (Winter 2009 Edition),* Edward N. Zalta, ed., http://plato.stanford.edu/archives/win2009/entries/moral-responsibility/.

Ewers, M., Sperling, R. A., Klunk, W. E., Weiner, M. W., and Hampel, H., 2011. Neuroimaging markers for the prediction and early diagnosis of Alzheimer's disease dementia. *Trends in Neurosciences,* **34**(8), pp. 430–442.

Faigman, D. L., 2013. Admissibility of neuroscientific expert testimony. In: S. J. Morse and A. L. Roskies, eds., *A Primer on Criminal Law and Neuroscience: A Contribution of the Law and Neuroscience Project, Supported by the MacArthur Foundation.* New York; Oxford: Oxford University Press, pp. 89–119.

Feresin, E., 2009. Lighter sentence for murderer with 'bad genes.' *Nature*, **10**(1038) / news.2009.1050.

Feresin, E., 2011. Italian court reduces murder sentence based on neuroimaging data. *Nature News Blog*, http://blogs.nature.com/news/2011/09/italian_court_reduces_murder_s.html.

Filiou, M. D., and Turck, C. W., 2011. General overview: Biomarkers in neuroscience research. *International Review of Neurobiology*, **101**, pp. 1–17.

Fischer, B. A., and Carpenter, W. T., Jr., 2009. Will the Kraepelinian dichotomy survive DSM-V? *Neuropsychopharmacology*, **34**(9), pp. 2081–2087.

Fischer, J., 1999. Recent work on moral responsibility. *Ethics*, **110**(1), pp. 93–139.

Fisher, R. S., Van Emde Boas, W., Blume, W., Elger, C., Genton, P., Lee, P., and Engel, J., Jr., 2005. Epileptic seizures and epilepsy: Definitions proposed by the International League Against Epilepsy (ILAE) and the International Bureau for Epilepsy (IBE). *Epilepsia*, **46**(4), pp. 470–472.

Forzano, F., Borry, P., Cambon-Thomsen, A., Hodgson, S. V., Tibben, A., De Vries, P., Van El, C., and Cornel, M., 2010. Italian appeal court: A genetic predisposition to commit murder. *European Journal of Human Genetics*, **18**(5), pp. 519–521.

Galanopoulou, A. S., 2010. Mutations affecting GABAergic signaling in seizures and epilepsy. *Pflügers Archiv—European Journal of Physiology*, **460**(2), pp. 505–523.

Gert, B., and Culver, C. M., 2004. Defining mental disorder. In: J. Radden, ed., *The Philosophy of Psychiatry: A Companion*. New York; Oxford: Oxford University Press, pp. 415–425.

Gert, B., Culver, C. M., and Clouser, K. D., 2006. Malady. *Bioethics: A Systematic Approach*. 2nd edition. New York; Oxford: Oxford University Press, pp. 129–164.

Gibson, G., 2012. Rare and common variants: Twenty arguments. *Nature Reviews Genetics*, **13**(2), pp. 135–145.

Glannon, W., 1998. Moral responsibility and personal identity. *American Philosophical Quarterly*, **35**(3), pp. 231–249.

Goddu, G. C., 2006. More on blameworthiness and alternative possibilities. *Journal of Moral Philosophy*, **3**(1), pp. 69–75.

Greene, J. A., 2007. *Prescribing by Numbers: Drugs and the Definition of Disease*. Baltimore, MD: Johns Hopkins University Press.

Hagerty, B., 2010. Can your genes make you murder? *National Public Radio* (July 1), http://www.npr.org/templates/story/story.php?storyId=128043329.

Halpern, M. T., Rnaud, J. M., and Vickrey, B. G. 2011. Impact of insurance status on access to care and out-of-pocket costs for U. S. individuals with epilepsy. *Epilepsy & Behavior*, **22**, 483-489.

Hanson, R. K., and Howard, P., 2010. Individual confidence intervals do not inform decision-makers about the accuracy of risk assessment evaluations. *Law and Human Behavior*, **34**, 275–281.

Hauser, W.A., and Mohr, J. P., 2011. Editorial: Seizures, epilepsy, and vascular malformations. *Neurology*, **76**(18), pp. 1540–1541.

Hibbeln, J. R., Ferguson, T. A., and Blasbalg, T. L., 2006. Omega-3 fatty acid deficiencies in neurodevelopment, aggression and autonomic dysregulation: Opportunities for intervention. *International Review of Psychiatry*, **18**(2), pp. 107–118.

Hope, T., Savulescu, J., and Hendrick, J., 2008. *Medical Ethics and Law: The Core Curriculum*. 2nd edition. Edinburgh: Churchill Livingstone.

Horwich, P., 1993. Bayesianism. In: J. Dancy and E. Sosa, eds., *A Companion to Epistemology*. Cambridge; Oxford: Blackwell, pp. 40–44.

Huang, C. J., and Baum, M. L., 2011. A rational basis for chemoprevention of prostate cancer. *American Journal of Bioethics*, **11**(12), pp. 27–29.

Huang, C. J., Parker, M., and Baum, M. L., 2014. Chapter 18, Genetics and dementia: Ethical concerns. In: C. Foster, J. Herring, and I. Doron (eds.), *Dementia: Law and Ethics*. Oxford: Hart, pp. 177–188.

Hyman, S. E., 2011. Diagnosing the DSM: Diagnostic classification needs fundamental reform. *Cerebrum*, http://dana.org/news/cerebrum/detail.aspx?id=32066.

Hyman, S. E., 2012. Revolution stalled. *Science Translational Medicine*, **4**, pp. 155cm11. See http://stm.sciencemag.org/content/4/155/155cm11.full-text.pdf+html

Hyötyläinen, T., 2012. Novel methodologies in metabolic profiling with a focus on molecular diagnostic applications. *Expert Review of Molecular Diagnostics*, **12**(5), pp. 527–538.

Insel, T. R., 2009. Translating scientific opportunity into public health impact: A strategic plan for research on mental illness. *Archives of General Psychiatry*, **66**(2), pp. 128–133.

Jack, C., Jr., and Holtzman, D., 2015. Biomarker modeling of Alzheimer's disease. *Neuron*, **80**, pp. 1347–1358.

Jackson, F., 1991. Decision-theoretic consequentialism and the nearest and dearest objection. *Ethics*, **101**(3), pp. 461–482.

Jaros, J. A. J., Guest, P. C., Ramoune, H., Rothermundt, M., Leweke, F. M., Martins-De-Souza, D., and Bahn, S., 2012. Clinical use of phosphorylated proteins in blood serum analysed by immobilised metal ion affinity chromatography and mass spectrometry. *Journal of Proteomics*, **76**, pp. 36–42.

John, S., 2011. Security, knowledge and well-being. *Journal of Moral Philosophy*, **8**(1), pp. 68–91.

Josephson, C. B., Leach, J.-., Duncan, R., Roberts, R. C., Counsell, C. E., and Salman, R. A., 2011. Seizure risk from cavernous or arteriovenous malformations: Prospective population-based study. *Neurology*, **76**(18), pp. 1548–1554.

Kamm, F. M., 2007. *Intricate Ethics: Rights, Responsibilities, and Permissible Harm*. Oxford: Oxford University Press.

Karlawish, J., 2010. Desktop medicine. *JAMA—Journal of the American Medical Association*, **304**(18), pp. 2061–2062.

Karlawish, J., 2011. Addressing the ethical, polic, and social challenges of preclinical Alzheimer disease. *Neurology*, **77**, pp. 1487-1493.

Karsten, S. L., Kudo, L. C., and Bragin, A. J., 2011. Use of peripheral blood transcriptome biomarkers for epilepsy prediction. *Neuroscience Letters*, **497**(3), pp. 213–217.

Klassen, T., Davis, C., Goldman, A., Burgess, D., Chen, T., Wheeler, D., Mcpherson, J., Bourquin, T., Lewis, L., Villasana, D., Morgan, M., Muzny, D., Gibbs, R., and Noebels, J., 2011. Exome sequencing of ion channel genes reveals complex profiles confounding personal risk assessment in epilepsy. *Cell*, **145**(7), pp. 1036–1048.

Kumar, V., Abbas, A., Fausto, N., and Aster, J., 2010. *Robbins and Cotran's Pathologic Basis of Disease*. 8th edition. Philadelphia; London: Saunders.

Kwan, J., 2010. Predicting the risk of poststroke epilepsy—why and how? *Nature Reviews Neurology*, **6**(10), pp. 532–533.

Lagnado, D. A., and Channon, S., 2008. Judgments of cause and blame: The effects of intentionality and foreseeability. *Cognition*, **108**(3), pp. 754–770.

Levy, N., 2011. *Hard Luck: How Luck Undermines Free Will and Moral Responsibility.* Oxford: Oxford University Press.

Lichtenstein, P., Yip, B. H., Björk, C., Pawitan, Y., Cannon, T. D., Sullivan, P. F., and Hultman, C. M., 2009. Common genetic determinants of schizophrenia and bipolar disorder in Swedish families: A population-based study. *The Lancet*, **373**(9659), pp. 234–239.

Luc Thiet Thuan v. R. [1997] AC 131, [1996] 2 All ER 1033, [1996] 3 WLR 45, [1996] 2 Cr App Rep 178, [1996] Crim LR 820, PC.

Malkesman, O., Austin, D. R., Chen, G. and Manji, H. K., 2009. Reverse translational strategies for developing animal models of bipolar disorder. *DMM Disease Models and Mechanisms*, **2**(5–6), pp. 238–245.

Martins-De-Souza, D., Guest, P. C., Vanattou-Saifoudine, N., Harris, L. W., and Bahn, S., 2011. Proteomic technologies for biomarker studies in psychiatry: Advances and needs. *International Review of Neurobiology*, **101**, 65–94.

Mattsson, N., Brax, D., and Zetterberg, H., 2010. To know or not to know: ethical issues related to early diagnosis of Alzheimer's disease. *International Journal of Alzheimer's Disease*, 2010, ID 841941, doi: 10.4061/2010/841941.

McKerlie, D., 1989. Equality and time. *Ethics*, **99**(3), pp. 475–491.

McKerlie, D., 1997. Priority and time. *Canadian Journal of Philosophy*, **27**(3), pp. 287–309.

Mitchell, K. J., 2011. The genetics of neurodevelopmental disease. *Current Opinion in Nurobiology*, **21**(1), pp. 197–203.

Mitchell, K. J., Huang, Z. J., Moghaddam, B., and Sawa, A., 2011. Following the genes: A framework for animal modeling of psychiatric disorders. *BMC Biology*, **9**, pp.76.

Monahan, J., 2014. The inclusion of biological risk factors in violence risk assessments. In: I. Singh, W. P. Sinnott-Armstrong, and J. Savulescu, eds., *Bioprediction, Biomarkers, and Bad Behavior: Scientific, Legal, and Ethical Challenges*. New York; Oxford: Oxford University Press, pp. 57–76.

Montmarquet, J., 1999. Zimmerman on culpable ignorance. *Ethics*, **109**(4), pp. 842–845.

Morse, S. J., 2013. Introduction. In: S. J. Morse and A. L. Roskies, eds., *A Primer on Criminal Law and Neuroscience: A Contribution of the Law and Neuroscience Project, Supported by the MacArthur Foundation*. New York; Oxford: Oxford University Press, pp. xv–xxiii.

Morse, S. J., and Newsome, W. T., 2013. In: S. J. Morse and A. L. Roskies, eds., *A Primer on Criminal Law and Neuroscience: A Contribution of the Law and Neuroscience Project, Supported by the MacArthur Foundation*. New York; Oxford: Oxford University Press, pp. 150–178.

Murphy, D., 2008. Concepts of disease and health. *The Stanford Encyclopedia of Philosophy*, Edward N. Zalta (ed.), http://plato.stanford.edu/archives/sum2009/entries/health-disease/.

Nelkin, D. K., 2011. George Sher, who knew? Responsibility without awareness. *Ethics*, **121**(3), pp. 675–680.

Nelson, B., and Yung, A. R., 2011. Should a risk syndrome for first episode psychosis be included in the DSM-5? *Current Opinion in Psychiatry*, **24**(2), pp. 128–133.

Nestler, E. J., Hyman, S. E., and Malenka, R. C., 2009. *Molecular Neuropharmacology: a Foundation for Clinical Neuroscience*. 2nd edition. New York; London: McGraw-Hill Medical.

Network and Pathway Analysis Subgroup of the Psychiatric Genomics Consortium, 2015. Psychiatric genome-wide association study analyses implicate neuronal, immune and histone pathways. *Nature Neuroscience*, **18**, pp. 199–209. doi: 10.1038/nn.3922.

Nussbaum, R. L., Mcinnes, R. R., and Willard, H. F., 2007. *Thompson & Thompson Genetics in Medicine*. 7th edition. Philadelphia; Edinburgh: Elsevier Saunders.

Olson, T. M., and Terzic, A., 2010. Human KATP channelopathies: Diseases of metabolic homeostasis. *Pflugers Archiv—European Journal of Physiology*, **460**(2), pp. 295–306.

Oriol, N. E., Cote, P. J., Vavasis, A. P., Bennet, J., Delorenzo, D., Blanc, P., and Kohane, I., 2009. Calculating the return on investment of mobile healthcare. *BMC Medicine*, **7**(27), doi: 10.1186/1741-7015-7-27.

Orr, S. P., Lasko, N. B., Macklin, M. L., Pineles, S. L., Chang, Y., and Pitman, R.K., 2012. Predicting post-trauma stress symptoms from pre-trauma psychophysiologic reactivity, personality traits and measures of psychopathology. *Biology of Mood & Anxiety Disorders*, **2**(1), pp. 8.

Oshana, M. A. L., 1997. Ascriptions of responsibility. *American Philosophical Quarterly*, **34**(1), pp. 71–83.

Padfield, N., 2008. *Criminal Law*. 6th edition. New York: Oxford University Press.

Pal, D. K., Pong, A. W., and Chung, W. K., 2010. Genetic evaluation and counseling for epilepsy. *Nature Reviews Neurology*, **6**(8), pp. 445–453.

Parfit, D., 1984. *Reasons and Persons*. Oxford: Oxford University Press.

Parfit, D., 1997. Equality and priority. *Ratio*, **10**(3), pp. 202–221.

Parfit, D., 2012. Another defence of the priority view. *Utilitas*, **24**(3), pp. 399–440.

Perkins, D., Jeffries, C., Addington, J., Bearden, C, Cadenhead, K., Cannon, T., Cornblatt, B., Mathalon, D., Mcglashan, T., Seidman, L., Tsuang, M., Walker, E., Woods, S., and Heinssen, R. 2015. Towards a psychosis risk blood diagnostic for persons experiencing high-risk symptoms: Preliminary results from the NAPLS project. *Schizophrenia Bulletin*,**41**(2), pp. 419–428.

Petrie, A. and Sabin, C., 2009. *Medical Statistics at a Glance*. 3rd edition. Chichester: Wiley-Blackwell.

Pies, R. W., 2010. Alzheimer's disease without dis-ease? New conundrums for psychiatric diagnosis and public health. *The Psychiatric Times*, July 19.

Pitkänen, A., Bolkvadze, T., and Immonen, R., 2011. Anti-epileptogenesis in rodent post-traumatic epilepsy models. *Neuroscience Letters*, **497**(3), pp. 163–171.

Polk News, 2009a. Waldroup gets 32 years. *Polk News* (May 13), http://www.polknewsonline.com/2009/05/13/Top_News/Waldroup_gets_32_years/4493.html.

Polk News, 2009b. Waldroup guilty, will not face death penalty. *Polk News* (March 25), http://www.polknewsonline.com/2009/03/25/Top_News/Waldroup_guilty,_will_not_face_death_penalty/4158.html.

Quehenberger, O., and Dennis, E. A., 2011. The human plasma lipidome. *New England Journal of Medicine*, **365**(19), pp. 1812–1823.

Rabinowicz, W., 2002. Prioritarianism for prospects. *Utilitas*, **14**(01), pp. 2–21.

Radden, J., 2002. Psychiatric ethics. *Bioethics*, **16**(5), pp. 397–411.

Ravizza, T., Balosso, S., and Vezzani, A., 2011. Inflammation and prevention of epileptogenesis. *Neuroscience Letters*, **497**(3), pp. 223–230.

Reid, R. C., 2012. From functional architecture to functional connectomics. *Neuron*, **75**(2), pp. 209–217.

Rice, D., 2014. D-Day: The most important weather forecast in history. *USA Today* (June 6), http://www.usatoday.com/story/weather/2014/06/03/d-day-weather-forecast/9914207/.

Roberts, J. S., and Tersegno, S. M., 2010. Estimating and disclosing the risk of developing Alzheimer's disease: Challenges, controversies and future directions. *Future Neurology*, **5**(4), pp. 501–517.

Ross, J., 2014. *The Forecast for D-Day: And the Weatherman behind Ike's Greatest Gamble*. Guildford, CT: Lyons Press.

Rousseau, J., 1979 [1762]. *Emile: or, On Education*. Basic Books. See https://books.google.com/books?id=VocWKgK9SxQC&printsec=frontcover&source=gbs_ge_summary_r&cad=0#v=onepage&q&f=false

Rutter, M., 2014. Biomarkers: Potential and challenges. In: I. Singh, W. P. Sinnott-Armstrong, and J. Savulescu, eds, *Bioprediction, Biomarkers, and Bad Behavior: Scientific, Legal, and Ethical Challenges*. New York; Oxford: Oxford University Press, pp. 188–205.

Sahakian, B. J., Malloch, G., and Kennard, C., 2010. A UK strategy for mental health and wellbeing. *The Lancet*, **375**(9729), pp. 1854–1855.

Sarnyai, Z., Alsaif, M., Bahn, S., Ernst, A., Guest, P.C., Hradetzky, E., Kluge, W., Stelzhammer, V., and Wesseling, H., 2011. Behavioral and molecular biomarkers in translational animal models for neuropsychiatric disorders. *International Review of Neurobiology*, **101**, pp. 203–238.

Savulescu, J. 2001 Procreative beneficence: Why we should select the best children. *Bioethics*, **15**(5-6), pp. 413–426.

Savulescu, J., and Kahane, G., 2011. Disability: A welfarist approach. *Clinical Ethics*, **6**(1), pp. 45–51.

Schizophrenia Working Group of the Psychiatric Genomics Consortium, 2014. Biological insights from 108 schizophrenia-associated genetic loci. *Nature*, **511**, pp. 421–427.

Scriver, C. R., and Waters, P. J., 1999. Monogenic traits are not simple: Lessons from phenylketonuria. *Trends in Genetics*, **15**(7), pp. 267–272.

Sebat, J., Levy, D. L., and McCarthy, S. E., 2009. Rare structural variants in schizophrenia: One disorder, multiple mutations; one mutation, multiple disorders. *Trends in Genetics*, **25**(12), pp. 528–535.

Shah, I., 1966. *The Exploits of the Incomparable Mulla Nasrudin*. Jonathan Cape.

Shah, I., 1971. *The Pleasantries of the Incredible Mulla Nasrudin*. New York: Dutton.

Shakespeare, W., [1608]1997. *King Lear.* New edn. Walton-on-Thames: Nelson.

Sher, G., 2006. Out of control. *Ethics*, **116**(2), pp. 285–301.

Sher, G., 2009. *Who knew? Responsibility Without Awareness.* Oxford: Oxford University Press.

Singh, I., and Sinnott-Armstrong, W. P., 2014. Introduction: Deviance, classification, and bioprediction. In: I. Singh, W. P. Sinnott-Armstrong, and J. Savulescu, eds., *Bioprediction, Biomarkers, and Bad Behavior: Scientific, Legal, and Ethical Challenges.* New York; Oxford: Oxford University Press, pp. 1–11.

Slobogin, C. 2014. Bioprediction in criminal cases. In: I. Singh, W. P. Sinnott-Armstrong, and J. Savulescu, eds., *Bioprediction, Biomarkers, and Bad Behavior: Scientific, Legal, and Ethical Challenges.* New York; Oxford: Oxford University Press, pp. 77–90.

Smith, H., 1983. Culpable ignorance. *The Philosophical Review*, **92**(4), pp. 543–571.

Smith (Morgan James) [2000] QB 1079, affd [2001] 1AC 146, [2000] 4 All ER 289, [2000] 3 WLR 654, [2001] 1 Cr App Rep 31, [2000] Crim LR 1004, HL.

Sperling, R. A., Aisen, P. S., Beckett, L. A., Bennett, D. A., Craft, S., Fagan, A. M., Iwatsubo, T., Jack Jr., C. R., Kaye, J., Montine, T. J., Park, D. C., Reiman, E. M., Rowe, C. C., Siemers, E., Stern, Y., Yaffe, K., Carrillo, M. C., Thies, B., Morrison-Bogorad, M., Wagster, M. V., and Phelps, C. H., 2011. Toward defining the preclinical stages of Alzheimer's disease: Recommendations from the National Institute on Aging-Alzheimer's Association workgroups on diagnostic guidelines for Alzheimer's disease. *Alzheimer's and Dementia*, **7**(3), pp. 280–292.

Stein, D. J., Phillips, K. A., Bolton, D., Fulford, K. W. M., Sadler, J. Z., and Kendler, K. S., 2010. What is a mental/psychiatric disorder? from DSM-IV to DSM-V. *Psychological Medicine*, **40**(11), pp. 1759–1765.

Steinlein, O.K., and Bertrand, D., 2010. Nicotinic receptor channelopathies and epilepsy. *Pflugers Archiv—European Journal of Physiology*, **460**(2), pp. 495–503.

Straussman, R., Morikawa, T., Shee, K., Barzily-Rokni, M., Qian, Z.R., Du, J., Davis, A., Mongare, M. M., Gould, J., Frederick, D. T., Cooper, Z. A., Chapman, P. B., Solit, D. B., Ribas, A., Lo, R. S., Flaherty, K. T., Ogino, S., Wargo, J. A., and Golub, T. R., 2012. Tumour micro-environment elicits innate resistance to RAF inhibitors through HGF secretion. *Nature*, **487**(7408), pp. 500–504.

Strawson, P.F., 2008. *Freedom and Resentment and Other Essays.* Oxford: Routledge.

Sung, J., Wang, Y., Chandrasekaran, S., Witten, D. M., and Price, N. D., 2012. Molecular signatures from omics data: From chaos to consensus. *Biotechnology Journal*, **7**(8), pp. 946–957.

Vickers, A. J., Basch, E., and Kattan, M. W., 2008. Against diagnosis. *Annals of Internal Medicine*, **149**(3), pp. 200–203.

Viding, E., and McCrory, E., 2014. Genetic biomarker research of callous-unemotional traits in children: Implications for the law and policymaking. In: I. Singh, W. P. Sinnott-Armstrong, and J. Savulescu, eds, *Bioprediction, Biomarkers, and Bad Behavior: Scientific, Legal, and Ethical Challenges.* New York; Oxford: Oxford University Press, pp. 153–172.

Vincent, N. A., 2013. Law and neuroscience: Historical context. In N. A. Vincent, ed, *Neuroscience and Legal Responsibility.* New York; Oxford: Oxford University Press, pp. 1–24.

Wakefield, J. C., 1992. Disorder as harmful dysfunction: A conceptual critique of DSM-III-R's definition of mental disorder. *Psychological Review*, **99**(2), pp. 232–247.

Wakefield, J. C., 2007. The concept of mental disorder: Diagnostic implications of the harmful dysfunction analysis. *World Psychiatry*, **6**(3), pp. 149–156.

Wallace, R. J., 1994. *Responsibility and the Moral Sentiments*. Cambridge, MA: Harvard University Press.

Wang, T. J., 2011. Assessing the role of circulating, genetic, and imaging biomarkers in cardiovascular risk prediction. *Circulation*, **123**, pp. 551–565.

Watson, G., 1996. Two faces of responsibility. *Philosophical Topics*, **24**, pp. 227–248.

Whiteford, H. A., Degenhardt, L., Rehm, J., Baxter, A. J., Ferrari, A. J., Erskine, H. E., Charlson, F. J., Norman, R. E., Flaxman, A. D., Johns, N., Burstein, R., Murray, C. J. L., and Vos, T., 2013. Global burden of disease attributable to mental and substance use disorders: Findings from the Global Burden of Disease Study 2010. *The Lancet*, **382**(9904), pp. 1575–1586.

Williams, B., 1973. A critique of utilitarianism. *Utilitarianism: For and Against*, **77**, pp. 108–118.

Williamson, J. R., Bliss, D. W., Browne, D. W., and Narayanan, J. T., 2012. Seizure prediction using EEG spatiotemporal correlation structure. *Epilepsy & Behavior*, **25**(2), pp. 230–238.

Wolpe, P. R, 2014. Rethinking the implications of discovering biomarkers for biologically based criminality. In: I. Singh, W. P. Sinnott-Armstrong, and J. Savulescu, eds, *Bioprediction, Biomarkers, and Bad Behavior: Scientific, Legal, and Ethical Challenges*. New York; Oxford: Oxford University Press, pp. 118–130.

Wood, P. L., 2012. Lipidomics of Alzheimer's disease: Current status. *Alzheimer's Research and Therapy*, **4**(1), 5.

Zhou, X., Maricque, B., Xie, M., Li, D., Sundaram, V., Martin, E. A., Koebbe, B. C., Nielsen, C., Hirst, M., Farnham, P., Kuhn, R. M., Zhu, J., Smirnov, I., Kent, W. J., Haussler, D., Madden, P. A. F., Costello, J. F., and Wang, T., 2011. The human epigenome browser at Washington University. *Nature Methods*, **8**(12), pp. 989–990.

Zimmerman, M. J., 1986. Negligence and moral responsibility. *Noûs*, **20**(2), pp. 199–218.

INDEX

www.ingramcontent.com/pod-product-compliance
Lightning Source LLC
LaVergne TN
LVHW011801200926
848484LV00003B/753
9780190236267